Digital Signal Processing for Audio Applications

Third Edition

Volume 2 – Code

Anton Kamenov

June 2017

ISBN-13: 978-0-692-91381-9

Foreword to the First Edition of Digital Signal Processing for Audio Applications

In the summer of 2003 we began designing multi-track recording and mixing software – Orinj at RecordingBlogs.com – a software application that will take digitally recorded audio tracks and will mix them into a complete song with all the needed audio production effects. Manipulating digital sound, as it turned out, was not easy. We had to find the answers of many questions, including what digital audio was, how we could mix audio tracks, how we could track the amplitude of digital sound so that we could apply compression, how we could track frequencies so that we could equalize, what a good model of artificial reverb would be, and many others. Bits of relevant information were available, albeit not always well organized and not always intuitive.

"Digital Signal Processing for Audio Applications" provides much of the needed information. It is a simple structured approach to understanding how digitally recorded sound can be manipulated. It presents and explains, and sometimes derives, the mathematical theory that the DSP user can employ in designing sound manipulating applications.

Although this book introduces much mathematics, we have designed it not for mathematicians, but for the engineers and hobbyists, who would be interested in the practical applications of DSP and not in its theoretical derivations. If properly explained, much of the practical DSP applications reduce to simple algebra. This said, we have included a sufficient amount of theory to provide an explanation of why DSP works the way it does. It is important for practitioners to have a good understanding of how DSP concepts come about. Much of the available DSP information has too much theory and not enough examples. Much of it has too many practical examples and not enough theoretical backing. We hope to have found the proper balance.

We hope you enjoy this book and make use of its definitions, explanations, and numerous examples.

The author and the administrators of www.recordingblogs.com

Foreword to The Code

Explaining the mathematics behind digital signal processing – DSP – is the task volume 1. It is a start, but there is more. It is not always straightforward to translate the mathematics into code. The purpose of volume 2 is just that. It translates the mathematical formulae in volume 1 into practical algorithms. It does so with actual DSP effects, including distortion, delay, chorus, equalizer, compressor, reverb, wah wah, and others.

Volume 1 of this book makes the argument that much of DSP can be reduced to simple algebraic and trigonometric manipulations. We hope that this volume shows that coding DSP is similarly not complex. In contemporary audio recording and mixing software, storing audio data, managing audio files, and designing an intuitive but functional user interface could be much more intricate than modifying the audio data themselves.

We hope you make use of this book and design some of your own DSP effects. They may just sound better than anyone else's. Audio production is as much an art, as it is science.

Many thanks to Mic of RecordingBlogs.com for providing access to the Orinj source code.

The author and the administrators of www.recordingblogs.com

Table of Contents

Table of Figures

Table of Code Samples

Chapter 1. Introduction

Each effect presented in this volume poses an inherently different problem that requires implementation specific to the effect. For example, how can one vary the time of chorus delays without producing discontinuities in the signal that would result in audible pops? How can one repeat the signal often enough to produce a continuous reverb rather than distinct delay repetitions if the number of repetitions in natural reverberations exceeds tens of millions? How can a vocal compressor reduce the amplitude of a signal quickly, but without changing the underlying wave form of the signal and therefore distorting the signal?

This said, there is also enough repeated and reusable code that allows us present effects in order of increasing complexity. All but one of the effects, for example, must look at past audio data and must therefore manage past audio data storage.

The code below uses Java, but the examples can easily be translated to other languages, as much of the code replicates mathematical formulae and uses basic data structures. Graphical user interfaces are the exception, but these are only discussed briefly for a couple of the effects.

Emphasis in this book is put on code that the reader can understand, rather than one that is most efficient. The approach to programming is a "brute force" one. It is not always elegant, but should be easy to replicate.

1.1. Reading this book

Chapter 2 of this book discusses the WAVE file format. Wave files are a common way to store audio data. If you are familiar with the format, you can safely skip over this chapter. In fact, reference to the information in chapter 2 is made only by chapter 5.

Chapter 3 presents the Orinj effect framework. We use this framework, as it provides ready code to test effects. You are not required to use it. We only explicitly refer to it in the first two DSP effects, but not afterwards. If you are developing audio effects for other applications and you have easy ways to test these effects, you can safely skip over this chapter as well.

Chapter 4 implements distortion. Certain types of distortion are the simplest digital signal processing effects. We present the full implementation of the effect and its graphical user interface in the Orinj effect framework and discuss undo, error checking, packaging, and obfuscating. We do so for the distortion only, but not for the remaining effects in the book. Only the effect implementations are included for those.

Chapter 5 presents the code for testing effects with the Orinj effect framework. If you are not planning to use this framework, you can skip over this chapter.

The remaining chapters of the book implement various effects. These chapters focus purely on the computation of the effect. They do not discuss its graphical user interface, testing, or packaging. Effects are presented in order of complexity: delay, echo, multitap delay, chorus, bass chorus, equalizer, noise gate, compressor, reverb, wah wah, and pitch shift. Note that some of these effects, such as the delay, are conceptually simple. They may have received little to no mention in volume 1 of this book. Others, such as the equalizer, are the subject of much of volume 1, but are only briefly covered here, as their implementation is not difficult.

Chapter 2. The WAVE file format

The wave file format is a widely-supported format for storing digital audio. A wave file uses the Resource Interchange File Format (RIFF) file structure and so data are organized in chunks as described below. Each chunk contains information about its type and size.

2.1. RIFF chunk

A wave file begins as follows.

Code 1. An example start of a WAVE file

```
0x52 0x49 0x46 0x46 0xss 0xss 0xss 0xss 0x57 0x41 0x56 0x45 ...
```

The first four bytes above are the ASCII characters RIFF. These bytes show that this is a RIFF file.

The next four bytes specify the size of the RIFF chunk in bytes. The size does not include the eight bytes for the characters RIFF and the size itself.

Each chunk in a RIFF file and all the chunks described above start with eight bytes, four of which determine the chunk type and four of which determine the chunk size. Since the size is always known, a software or a device that must interpret a RIFF file does not have to understand all chunks. It can skip over those chunks that it does not understand.

The next four bytes in the example above are the ASCII characters WAVE. These show that this RIFF file is, in fact, a wave file.

2.2. Wave chunks

The twelve bytes in the example above are followed by chunks of information. A wave file can have various types of chunks, some of which provide additional detail on the format of the file, others contain audio data, and still others contain meta data, such as markers and cues. The most common chunks are described below. Others are listed in Appendix A.

A wave file always contains at least a format chunk and a data chunk, in no particular order. It does not have to contain any of the other chunks. It may also contain chunks that most software or devices do not understand, such as chunks designed by certain software producers for their software.

2.3. Endianism

All information is stored with the least significant byte first (little-endian). For example, if the size is contained in the four bytes **0x88 0x58 0x01 0x00** in this order, then the size is the hexadecimal value 0x00015888 bytes (decimal value 88,200).

2.4. Word alignment

All information in a wave file must be word aligned (i.e., aligned at every two bytes). If a chunk has an odd number of bytes, then it is padded with a zero byte, although this byte is not counted in the size of the chunk.

2.5. Format chunk

The format chunk in the Wave file format has the following structure.

Figure 1. Structure of the format chunk in a wave file

Description	Length in bytes	Starts at byte in the chunk	Value
chunk ID	4	0x00	The ASCII character string "fmt " (note the space at the end)
size	4	0x04	The size of the format chunk
compression code	2	0x08	Various
number of channels	2	0x0A	Various
sampling rate	4	0x0C	Various
average bytes per second	4	0x10	Various
block align	2	0x14	Various
significant bits per sample	2	0x16	Various
number of extra format bytes	2	0x18	Various
extra format bytes	various	0x1A	Various

chunk ID – The chunk ID is always "fmt " (with a space at the end, as all chunk IDs have four bytes). This chunk ID shows that this is a format chunk.

size – As always, the size of the chunk is the size of the data that follow the chunk ID and the size itself. The typical size of the format chunk is 16 bytes, but the format chunk could be larger, if there are extra format bytes (the last two rows of the table above).

compression code – There are over 100 different compression codes and perhaps even over 200. One common compression code is 1, for Microsoft PCM uncompressed data. PCM, or pulse code modulation, is described in chapter 3 of volume 1. It is the process by which analog sound data are sampled at uniform intervals and the samples are recorded with a uniform scaling. In other words, PCM uses a uniform sampling rate and uniform sampling resolution. With compression code 1, the sample values are stored as signed or unsigned integers as discussed below. This is the compression code assumed in the rest of this book.

Another common compression code is 3, or the Microsoft IEEE float. Sample values are stored similarly, but as floating-point numbers, rather than as integers. Compression codes 6 (ITU G.711 A-law) and 7 (ITU G.711 μ-law) are also commonly used, typically in telephone systems and early browsers, usually to compress 8-bit PCM recordings. Unlike the Microsoft PCM and IEEE float compressions, ITU G.711 A-law ITU G.711 μ-law compress the dynamic range of the signal and result in some loss of information.

number of channels – The typical number of channels is 1 in mono waves and 2 in stereo waves. There can be other values. Quadrophonic sound is one type of surround sound that uses four channels. Typical home theater setups use six or eight channels, typically denoted as 5.1 or 7.1 surround sound. In this book, we work almost exclusively with mono waves, although we show examples that handle stereo audio.

sampling rate – The sampling rate is the number of samples per second. CD quality audio, for example, uses 44,000 Hz and the corresponding value recorded here is 44100. As discussed in volume 1 of this book, larger sampling rates produce better quality audio as they more closely represent the signal. Contemporary audio recording may use higher sampling rates, such as 96,000 Hz. This book discusses exclusively 44100 Hz audio, but its examples do not require coding changes if used on other sampling rates.

average bytes per second – An uncompressed PCM wave file that has a sampling rate of 44100 Hz, 1 channel, and sampling resolution of 16 bits (2 bytes) per sample, for example, has an average number of bytes equal to 44100 * 2 * 1 = 88,200 per second. Certain types of compression may reduce the number of bytes stored in the wave file per second and may do so based on the actual values of the signal, resulting in different numbers of bytes at different signal times. Thus, this value is only an average and not a precise constant.

block align – This is the total size of a sample in the wave file. For example, a PCM wave that has a sampling resolution of 16 bits (2 bytes) and 2 channels records a block of samples in 2 * 2 = 4 bytes.

significant bits per sample – This number is the sampling resolution of the file. It is the number of bits used to record a sample per channel. If a sample uses 2 bytes or 16 bits, this value is 16. A typical sampling resolution is 16 bits per sample, but could be anything greater than 1. Common sampling resolutions are 8, 16, 24, and 32. In uncompressed, integer PCM format, the sampling resolution determines the number of values that a signal can take. For example, with 16 bits, there can be at most $2^{16} = 65,536$ values. Since a signal oscillates, say, taking positive and negative values, the maximum peak amplitude that can be recorded is $2^{15} = 32,768$ (technically, -32,768 to 32,767). The maximum peak amplitude with 8-bit recording on the other hand is 128, which implies that the signal itself may take much fewer values than 16-bit recording and is therefore more imprecise. In this book, we work with 16-bit audio, although we do present code snippets for interpreting other sampling resolutions.

number of extra format bytes – This field may or may not be present. It determines the number of extra bytes that follow.

extra format bytes – These also may or may not be present. These typically are not present in uncompressed PCM files, such as the ones discussed in this book, as in these files there is no need to include additional information about the file format.

2.6. Data chunk

The data chunk in the Wave file format has the following structure.

Figure 2. Structure of the data chunk in a wave file

Byte sequence description	Length in bytes	Starts at byte in the chunk	Value
chunk ID	4	0x00	The ASCII character string "data"
size	4	0x04	The size of the data chunk (number of bytes) less 8 (less the "chunk ID" and the "size")
data	various	0x08	The sampled audio data

chunk ID – The chunk ID is always the ASCII string "data", signifying that this is a data chunk.

size – The size of the chunk is the size of the data that follow the chunk ID and the size itself. This is also the size of the sampled audio data. For example, one second audio recorded at 44,100 Hz on 1 channel with 16 bits sampling resolution has 44100 * 1 * 16 / 8 = 88,200 bytes of data.

data – This is the portion of the wave file that contains the actual sampled audio data.

How samples are stored depends on the format specified in the format chunk. This is explained with the example below.

2.7. An example of an actual wave file

Consider the following sequence of bytes, taken from the start of an actual wave file.

Code 2. Contents of an example wave file

```
    0x52 0x49 0x46 0x46 0x24 0xA0 0xAA 0x00 0x57 0x41 0x56 0x45 0x66 0x6D 0x74
0x20 0x10 0x00 0x00 0x00 0x01 0x00 0x02 0x00 0x44 0xAC 0x00 0x00 0x10 0xB1 0x02
0x00 0x04 0x00 0x10 0x00 0x64 0x61 0x74 0x61 0x00 0xA0 0xAA 0x00 0x00 0x00 0x00
0x00 0x00 0x00 0x00 ...
```

This sequence of bytes represents the following.

0x52 0x49 0x46 0x46 – This is the ASCII character string "RIFF", which means that this is a RIFF file.

0x24 0xA0 0xAA 0x00 – These bytes form the four-byte value 0x00AAA024, which is the decimal value 11182116. This means that the size of the file is 11182116 bytes excluding the eight bytes for the RIFF string and the size itself. The total file size is 11182116 + 8 = 11182124.

0x57 0x41 0x56 0x45 – This is the ASCII character string "WAVE" and so the file is a WAVE RIFF file.

0x66 0x6D 0x74 0x20 – This is the ASCII identification of the first chunk in the wave portion of the RIFF file. In this case, the ID is the ASCII character string "fmt ", which means that this is the format chunk. In this file, the format chunk happens to be the first chunk, although this is not always the case.

0x10 0x00 0x00 0x00 – These bytes form the hexadecimal value 0x00000010, which is the decimal value 16. The format chunk has 16 bytes after its ID and size. There are no extra bytes in this format chunk.

0x01 0x00 – The compression code in the format chunk is 0x0001 (decimal value 1) and so this file contains uncompressed PCM data. The audio data have been sampled with a constant, uniform sampling rate and recorded with a uniform sampling resolution.

0x02 0x00 – There are 0x0002 (decimal value 2) channels in the audio in this file.

0x44 0xAC 0x00 0x00 – These four bytes form the hexadecimal value 0x0000AC44, which is the decimal value 44100. The sampling rate of this file is 44100 Hz. There are 44100 samples per channel for each second of audio.

0x10 0xB1 0x02 0x00 – The average number of bytes per second is 0x0002B110 or 176400. Note below that each sample is recorded in two bytes and, as above, there are 44100 samples for each of the two channels. Thus, 176400 = 44100 * 2 * 2. Since this is an uncompressed PCM file and the sampling rate and resolution are constant, the average number of bytes per second is also the actual number of bytes per second for each second of audio.

0x04 0x00 – The sample at each point of time is recorded in 0x0004 (decimal value 4) bytes. There are two channels and each sample for each channel is recorded in two bytes (see below). Thus, the block align is 4 and, in this uncompressed PCM file, one can move through the sampling points of time by reading the audio data four bytes at a time.

0x10 0x00 – The number of significant bits per sample is 16. Thus, each sample is recorded in 16 bits or 2 bytes. This is the sampling resolution of the audio. This is the end of the format chunk, since it amounts to a total of 16 bytes, which was the size of this chunk. There were two bytes for the compression code, two bytes for the number of channels, four bytes for the sampling rate, four bytes for the average number of bytes per second, two bytes for the block align, and two bytes for the sampling resolution. The bytes that follow should represent another chunk of the wave file. These bytes are as follows.

0x64 0x61 0x74 0x61 – This is the ASCII character string "data", which means that the next chunk is the data chunk.

0x00 0xA0 0xAA 0x00 – The size of the data chunk (minus the ASCII string "data" and the size itself), is 0x00AAA000 or 11182080. At this point, it is known that the file contains 11182080 / (2 channels * 2 bytes per sample) = 2795520 samples per channel and that it is approximately 2795520 / 44100 = 63.39 seconds.

0x00 0x00 – This is the first sample for channel 1. It is not surprising that its value is zero. Recordings often start with a brief period of silence.

0x00 0x00 – This is the first sample for channel 2.

0x01 0x00 – This is the second sample for channel 1. Its value is 0x0001, which is 1 decimal. Since this recording uses 16 bits per sample, sample values can be between $\pm 2^{15} = \pm 32768$. The value of this sample is quite low. It is $1 / 32768 \approx 0.00003$ of the maximum signal value that can be recorded or $20 \log_{10}(1 / 32768) \approx -90.3$ decibels. Note that the difference between the maximum and minimum amplitude that the human ear can perceive is approximately 90 decibels, which explains the choice of the 16-bit sampling resolution in CD-quality audio.

0x00 0x00 – This is the second sample for channel 2.

. . . – And so on.

The order of samples in stereo data is as follows.

Figure 3. Order of samples in stereo WAVE data

```
sample 0 for channel 1
sample 0 for channel 2
sample 1 for channel 1
sample 1 for channel 2
sample 2 for channel 1
sample 2 for channel 2
. . .
```

By convention, although different data representations are possible, when the sampling resolution of uncompressed PCM files (compression code 1) is 8-bit, the audio samples in the data chunk use unsigned integer numbers. Thus, audio samples range between 0 and 255 (0x00 and 0xFF) and silence is 127 (0x7F). When the sampling resolution of uncompressed PCM files is 16-bit, 24-bit, or 32-bit, the audio samples in the data chunk are signed integers. Silence then is represented by 0.

2.8. Other wave chunks

This book does not list all wave chunks and does not use chunks other than the ones described above. Other possible chunks, but not all, are listed in Appendix A.

Since the RIFF format is used for other types of files, such as AVI files, a RIFF file can also contain types of chunks that are not relevant to the wave file format. For example, the junk and pad chunks are used to add random data to the file to, perhaps, align the file chunks on a 2K boundary. A software application does not have to recognize or use all chunk types and may ignore certain chunks. Since the sizes of chunks are known, applications can simply skip over chunks that they do not recognize.

> The code in this book assumes that the wave files are uncompressed PCM wave files that contain a format chunk and one data chunk. They use a sampling resolution of 16 bits, the sampling rate 44100 Hz, and one audio channel. Modifying the code for different sampling resolutions, sampling rates, and channels is relatively simple. In addition, Java has standard tools to convert files with certain other compression codes to uncompressed PCM waves.

Chapter 3. The Orinj effect framework

Orinj uses a strict framework for effects. Effects must implement specific Java interfaces and extend specific Java classes. Effects must also be placed in Java JAR files that contain XML descriptions according to a specific XML schema. This allows Orinj to recognize and import effects during startup that are not a part of the Orinj code and installation. Effects can be created and updated without changes to Orinj. Similarly, Orinj can be updated without changing the underlying effects. Anyone can create effects for Orinj.

The Orinj framework is used in this book mainly as it provides for an easy testing environment. Effects can be tested in actual Orinj sessions. Alternatively, RecordingBlogs.com contains ready source code for testing effects with the Orinj effect framework outside of Orinj.

3.1. oreffect.jar

oreffect.jar is a Java JAR file and a part of the Orinj effect framework. It is a part of the Orinj installation and resides in the "Orinj/orange" folder of the installation. It and its source can also be downloaded directly from RecordingBlogs.com.

This package specifies the two interfaces that an Orinj effect should implement in Java – **EffectInterface** and **EffectPanelInterface**. When you design Orinj effects, you will at a minimum develop two Java classes – one of the effect and one for the graphical user interface of the effect. The graphical user interface contains the controls that the user can change to modify the effect. These two classes must implement the two interfaces noted here respectively.

EffectInterface – This interface is implemented by the actual effect. For example, Orinj effects that implement this interface contain a function **apply** with a specific syntax, so that Orinj can send buffers with audio data to these effects. Examples are below.

EffectPanelInterface – This interface is implemented by the graphical user interface for the effect. For example, the graphical user interfaces for the Orinj effects contain the function **updateData** so that Orinj can tell the effect when its user controls should be updated with the effect values.

The remaining Java classes in oreffect.jar are as follows. The source code for each class and further explanations are included below.

EffectFont – This class contains two static fonts of class **java.awt.Font**. You can use these fonts for the graphical user interface for your effect, so that the look and feel of your effect is the same as the look and feel of the rest of Orinj. You are not required to use these fonts.

Undo – All effects (the actual effect and not its graphical user interface) should extend this class. This class contains an array of event listeners and can be used to fire undo events, which Orinj uses to store undo information for your effect. With this, the Orinj user can undo changes made to the controls in your effect. The effect must extend this class even if it does nothing further to implement undo.

UndoEvent – The event that should be fired so that Orinj stores undo information and displays an undo message in its menus.

UndoListener – An interface that is implemented by the listener for undo events. This interface is implemented by components of Orinj, who listen for events fired by effects.

All classes in oreffect.jar are described below in alphabetical order.

3.2. EffectFont.java

This class simply contains two statically defined fonts of the class **java.awt.Font**. You are not required to use these fonts in your effects.

Code 3. EffectFont.java

```
package recordingblogs.com.oreffect;

import java.awt.Font;

public class EffectFont
{
    public static final Font LARGEFONT = new Font("SansSerif", 0, 12);
    public static final Font SMALLFONT = new Font("Arial", 0, 9);
}
```

Font LARGEFONT – In the Orinj effects that are distributed with the Orinj installation, this font is used primarily for control labels, buttons, and text fields.

Font SMALLFONT – In the Orinj effects that are distributed with the Orinj installation, this font is used primarily for slider labels.

3.3. EffectInterface.java

All Orinj effects must implement this class.

Code 4. EffectInterface.java

```
package recordingblogs.com.oreffect;

import javax.sound.sampled.AudioFormat;

public interface EffectInterface
{
    public abstract void apply(byte [] drybuffer, byte [] wetbuffer,
        byte [] controlbuffer, AudioFormat format, double time);
    public abstract boolean allowsDryWetMix();
    public abstract boolean allowsSideChaining();
    public abstract void startPlay();
    public abstract void stopPlay();
    public abstract boolean hasData();
    public abstract boolean readObject(ReadInterface stream);
```

```
    public abstract boolean writeObject(WriteInterface stream);
    public abstract void setEqual(EffectInterface effect);
    public abstract void setLanguage(String languageCode);
}
```

The following is a description of the member functions of this interface.

void apply(byte [] drybuffer, byte [] wetbuffer, byte [] controlbuffer, AudioFormat format, double time) – This function applies the effect to the sound data that are contained in **drybuffer**.

- **byte [] drybuffer** – This buffer contains the incoming audio data. The effect is applied to these data. This is an array of bytes and must be translated into sampled audio data depending on the argument **format**. This book assumes that **drybuffer** contains 16-bit audio of a single channel recorded with the 44100 Hz sampling rate. In Orinj, **drybuffer** may contain: unsigned 8-bit integer data, signed 16-bit integer data, signed 24-bit integer data, or signed 32-bit floating-point data. There may be one or two channels. The sampling rate may also be different.

  drybuffer does not contain all audio data that are sent to the effect. As playback progresses, audio data are sent to the effect in pieces. It is up to the effect to store, keep, and discard previous audio data as needed. Most effects must look at previous audio data buffers. An echo must do so, as the current value of the echo repetitions depends on the past values of the signal. The echo effect implemented later in this book stores previous audio data, uses them, and discards them when they are no longer necessary (i.e., when the audio data are too far in the past to be used by the echo).

- **byte [] wetbuffer** – This buffer should be treated as an empty one and the output of the effect should be placed in this it. If the effect has a *dry* and *wet* mix as explained below (a mix between the original signal and any changes or additions introduced by the effect), then place only the wet data into this buffer. If your effect does not have a dry and wet mix, place the total effect output in this buffer.

  Consider the Orinj Delay. This effect takes the original signal and creates one delayed and decayed repetition of that signal. The dry audio data are the original signal. The wet audio data are the repetition. The simple delay effect in Orinj leaves **drybuffer** unchanged, but places the repetition in **wetbuffer**. Since Orinj knows that this effect has a dry and wet mix (see **allowsDryWetMix** below), Orinj will use both the dry and wet buffers in further processing.

  Consider the Orinj Compressor. This effect does not have a dry and wet mix. It leaves **drybuffer** unchanged, but places the compressed signal in **wetbuffer**. Since Orinj knows that this effect does not have a dry and wet mix, it will ignore **drybuffer** coming out of this effect in further processing.

The length of **wetbuffer** is the same as the length of **drybuffer**. The format of audio data in **wetbuffer** should be the same as the format of audio data in **drybuffer**. This format is defined by the argument **format**.

- **byte [] controlbuffer** – This buffer is used by effects that allow side chaining (see **allowsSideChaining** below). The Orinj Side chained compressor adjusts the levels of a signal depending on the levels of another signal (i.e., of one track depending on another in the Orinj multitrack session view). In this effect, **controlbuffer** contains the audio data of the second track – the one whose amplitudes are considered in adjusting the amplitudes of the first track.

  The length of **controlbuffer** is the same as the length of **drybuffer**. The format of audio data in **controlbuffer** is the same as the format of audio data in **drybuffer**. **controlbuffer** may be null or empty for effects that do not specify that they can use side chaining.

- **AudioFormat format** – This is the format of audio data contained in **drybuffer**, **wetbuffer**, and **controlbuffer**. The class **javax.sound.sampled.AudioFormat** contains much of the same information as the format chunk of a wave file.

- **double time** – This is the start time in seconds of the audio buffers (e.g., **drybuffer**) in the session or wave that is being played. For example, if playback starts at 10 seconds into the multitrack session, **time** is equal to 10 in the first call to this function.

A choice was made in Orinj effects to use unaltered audio data, passed to the effect as byte arrays. This is different than in other music recording software, which may convert these data to floating-point or other arrays. This means that the effects in Orinj themselves must interpret the data based on their audio format. Examples of such data conversion are in chapter 4.

The argument **time** is only used by effects with audible changes in parameters in time. Only the wah wah in chapter 15 of this book uses this parameter, as the position of the wah wah filter depends on the current playback time and shifts back and forth in the frequency spectrum during playback. The chorus and bass chorus of chapters 9 and 10 also have parameters that change in time – the delays between the original signal and chorus repetitions – but such changes and not easily perceived by the listener and hence the chorus effects do not use this argument.

boolean allowsDryWetMix() – This function simply returns **true** if the effect allows dry and wet mix and **false** if it does not. All effects place computed audio data in the wet buffer and leave the dry buffer unchanged. If an effect allows dry and wet mix, Orinj will use both the original dry signal and the computed wet signal and can adjust the mix between these two separate signals outside of the effect itself. If an effect does not allow dry and wet mix, Orinj will use only the wet buffer and ignore the dry buffer after the effect.

boolean allowsSideChaining() – This function returns **true** if the effect allows side chaining and **false** if it does not. A side chained compressor, for example, adjusts one track depending on the information of another track (the control track). The intent is for the application to provide buffers with sound data to the effect for both tracks – the one to be changed and the one that controls. If the effect does not allow side chaining, the control buffer in the **apply** function may be null and should be ignored.

void startPlay() – This function implements any preparations that must occur at the start of playback. The Orinj Graphic equalizer, for example, implements this function to compute its frequency filters, although these filters may be recomputed again during playback if the user changes the equalizer controls.

void stopPlay() – This function should implement any cleanup actions that must take place at the stop of playback. This function must be implemented carefully, as playback may continue for a small amount of time even after this function is called, depending on the size of audio buffers in Orinj. This function, for example, must not remove stored incoming audio data. Rather, these data can be removed at the start of the next playback. Very few effects distributed with the Orinj installation use this function.

boolean hasData() – This function returns **true** if the effect has not finished processing and **false** otherwise. Take a simple delay that creates a single repetition of the original signal. Even if the original signal has ended (e.g., at the end of the last wave in the track), the repetition of the signal created by the effect may continue. During playback, Orinj checks whether there is original signal. If not, it asks the effect whether the effect itself will produce a signal. If yes, the processing of the effect continues. If not, the effect stops. This saves computations in Orinj making playback and mixing faster.

To create its repetition, a simple delay stores a portion of the original signal. After the original signal is finished, the delay returns **true** with this function only if it still has a portion of the original signal stored. In this case, it can still produce the repeated signal. When the saved portion of the original signal is also finished, the delay will no longer produce the repeated signal and **hasData** returns **false**.

It is always safe to just return **false**. If the effect does so, however, the wet (processed) signal from the effect may stop prematurely at the end of a wave. Most effects, in fact, delay the signal and that delay should be accounted for.

boolean readObject(ReadInterface stream) – This function and the next function allow Orinj to save the effect controls in the session files, loop files, or for undo. Orinj expects that each object (tracks, effects, volume envelopes) implements its own serialization methods. Examples are below. The standard Java serialization via **Serializable** is not used. The class **ReadInterface** is a part of oreffect.jar and is described below.

In addition to effect controls, effects can also store and read a version number that is independent of the Orinj version. In this way, effects can implement version control.

boolean writeObject(WriteInterface stream) – This is the function that Orinj uses to save effects in the Orinj sessions, loops, or for undo. This function is implemented similarly to the one above, but values are stored rather than read. **WriteInterface** is a part of oreffect.jar and is similarly described below.

void setEqual(EffectInterface effect) – This function sets the member data of the effect equal to the member data of the argument **effect**. The implementation should check whether the two effects are of the same class. This function is not used in the current version of Orinj, but may be used in future versions.

void setLanguage(String languageCode) – This function is implemented if the graphical user interface for the effect supports different languages. This function changes the current language to the one specified by **languageCode**. **languageCode** is the three letter ISO 639-2 language.

The implementation of languages is not discussed in this book. Orinj implements languages using XML files with the XML schema orinjlang.xsd, which can be downloaded from RecordingBlogs.com. However, designers are free to implement languages in any way they see fit. If languages or the language code sent to the effect with this function are not supported by the effect, the effect can simply use a default language.

3.4. EffectPanelInterface.java

Code 5. EffectPanelInterface.java

```
package recordingblogs.com.oreffect;

public interface EffectPanelInterface
{
    public abstract void updateData();
}
```

The graphical user interface for Orinj effects is an extension of the class **javax.swing.JPanel** and implements this interface.

In Orinj, all effect dialogs are standard. The actual dialog (of class **JDialog**) is already implemented. The graphical user interfaces for the effects do not extend **JDialog**, but **JPanel**, and are simply added to the dialog.

The effect dialog that encompasses the effect **JPanel** has: a Close button that closes the dialog; a bypass check box that can be clicked so that the effect is bypassed and not used during Orinj playback or mixing; a text field to change the name of the effect; a drop-down box to select presets, which are predefined values for each of the effect controls; buttons to save or delete presets; and a drop-down box to select a control track, if the effect allows side chaining. These controls are already implemented and should not be recreated. Only effect specific controls should be implemented. For the Orinj delay, for example, these specific controls are the left and right channel delays, decays, and polarity.

EffectPanelInterface contains a single function. Examples of how this function is implemented are below.

void updateData() – This function should set the values of the effect panel controls to the values of the member data of the effect itself.

3.5. ReadInterface.java

Effects do not implement this interface, but they use objects that implement this interface.

Code 6. ReadInterface.java

```java
package recordingblogs.com.oreffect;

import java.io.IOException;

public interface ReadInterface
{
    public boolean readBoolean() throws IOException;
    public byte readByte() throws IOException;
    public short readShort() throws IOException;
    public int readInt() throws IOException;
    public long readLong() throws IOException;
    public float readFloat() throws IOException;
    public double readDouble() throws IOException;
    public String readString() throws IOException, ClassNotFoundException;
    public int read(byte [] value) throws IOException;
}
```

The functions in this class do precisely what their names suggest. Note that the argument **value** of the last function in this interface cannot be null.

There are two classes in Orinj that implement this interface.

- The first class is an extension of **java.io.ObjectInputStream**. Most of the functions of **ReadInterface** are, in fact, the functions of **ObjectInputStream**. This class and its functions are used in the exact same way one would use **ObjectInputStream** and the functions that belong to that class. Examples are below. In Orinj, **ReadInterface** reads the member data of the effect as part of the serialization of the session or loop.
- The second class in Orinj that implements this interface reads the member data of the effect when the user undoes changes in the effect. This class is implemented very differently from **ObjectInputStream**. It does not read information from disk, but from memory.

Having these two classes implement the same interface means that one only needs to implement one **readObject** function for each effect. This function is called both when the effect is read as part of a session or loop and when this effect is read for the purposes of undo.

3.6. Undo.java

This class should be extended by all effects, even if these effects to do not actually implement undo (i.e., even if they do not fire undo events). This class essentially contains an array of event listeners, which are notified when an undo event is fired by the effect.

Code 7. Undo.java

```java
package recordingblogs.com.oreffect;

import javax.swing.event.EventListenerList;

public class Undo
{
    private EventListenerList m_listeners;

    public Undo()
    {
        m_listeners = new EventListenerList();
    }

    public void addUndoListener(UndoListener listener)
    {
        m_listeners.add(UndoListener.class, listener);
    }

    public void removeUndoListener(UndoListener listener)
    {
        m_listeners.remove(UndoListener.class, listener);
    }

    protected void fireUndoEvent(UndoEvent event)
    {
        Object[] listeners = m_listeners.getListenerList();
        for (int i = 0; i < listeners.length; i += 2)
        {
            if (listeners[i + 1] instanceof UndoListener)
                ((UndoListener) listeners[i + 1]).undoStorageRequired(event);
        }
    }
}
```

The following are the members of the class.

m_listeners – the array of undo event listeners.

Undo() – the default and only constructor.

void addUndoListener(UndoListener listener) – This function adds a listener to the array of event listeners in the effect. This function is used by Orinj to add its components as event listeners.

void removeUndoListener(UndoListener listener) – This function removes a listener from the array of event listeners in the effect. This function is also used by Orinj.

void fireUndoEvent(UndoEvent event) – This function notifies all listeners of the undo event **event**. This function is used by the effect to fire undo events when the effect controls change. Examples are below. The undo events will notify the listeners (the Orinj components) that the effect is changing and its data should be stored so that they can be brought back if the user wants to do so. The undo events will also provide Orinj with the message to be displayed in the undo menu.

3.7. UndoEvent.java

Code 8. UndoEvent.java

```java
package recordingblogs.com.oreffect;

import java.util.EventObject;

public class UndoEvent extends EventObject
{
    private String m_undoMessage;

    public UndoEvent(Object source, String undoMessage)
    {
        super(source);
        m_undoMessage = new String(undoMessage);
    }

    public String getUndoMessage()
    {
        return m_undoMessage;
    }
}
```

This is the undo event that should be fired by effects through the Undo class extended by the events, to notify Orinj that the effect has changed and to provide Orinj with the message that should be placed in the undo menu. This class simply contains one member data item – the message – and one member function besides the constructor – the function that provides access to the message.

String m_undoMessage – The undo message is the message that will be displayed in the Orinj menu. For example, if the user makes a change to the left channel decay in the Orinj Delay effect, the message might say "Undo Left Channel Decay".

UndoEvent(Object source, String undoMessage) – This is the default and only constructor. **source** is the object that fires the undo message. **undoMessage** is the undo message that will be stored in **m_undoMessage**.

String getUndoMessage() – This function provides access to the undo message.

3.8. UndoListener.java

Code 9. UndoListener.java

```
package recordingblogs.com.oreffect;

import java.util.EventListener;

public interface UndoListener extends EventListener
{
    public void undoStorageRequired(UndoEvent e);
}
```

This is the listener for undo events. This interface is used by Orinj, where the listeners for undo events reside. It is not used by effects. It is implemented by the dialog that contains the **JPanel** graphical user interface for the effects with controls for the effects.

void undoStorageRequired(UndoEvent e) – This function stores the undo information and sets the text of the undo menu to the appropriate undo message.

3.9. WriteInterface.java

Code 10. WriteInterface.java

```
package recordingblogs.com.oreffect;

import java.io.IOException;

public interface WriteInterface
{
    public void writeBoolean(boolean value) throws IOException;
    public void writeByte(int value) throws IOException;
    public void writeShort(int value) throws IOException;
    public void writeInt(int value) throws IOException;
    public void writeLong(long value) throws IOException;
    public void writeFloat(float value) throws IOException;
    public void writeDouble(double value) throws IOException;
    public void writeString(String value) throws IOException;
    public void write(byte [] value) throws IOException;
}
```

As with **ReadInterface**, there are two classes in Orinj that implement this interface. The first class is an extension of **java.io.ObjectOutputStream** and is used to save sessions and loops (and the effects as part of sessions and loops) for serialization. The second class saves effect information for undo.

Chapter 4. Distortion

In music, *distortion* is the sound effect that occurs when the amplitude peaks of a signal are compressed or cut off when the audio equipment is overloaded. Chapter 6 of volume 1 contains the mathematics behind the distortion effect described below.

There are various types of distortion. The signal peaks may be cut off, which is called a *clip* or a *hard clip*. The signal peaks may be compressed and then cut off, which is called a *soft clip*. The signal peaks may also be simply compressed. These are the three types of distortion implemented in this chapter.

4.1. Implementation of the distortion

The following is an implementation of the simple delay in the Orinj framework, assuming one channel of audio with the sampling resolution of 16 bits. The graphical user interface for this delay is described in the sections that follow.

Code 11. Distortion

```java
package mycompany.com;

import java.io.*;
import javax.sound.sampled.AudioFormat;
import recordingblogs.com.oreffect.*;

public class Distortion extends Undo implements EffectInterface
{
   public static final String [] TYPES = {"Hard clip", "Soft clip", "No clip"};

   public static final int HARDCLIP = 0;
   public static final int SOFTCLIP = 1;
   public static final int NOCLIP = 2;

   public static final float MINTHRESHOLD = -50F;
   public static final float MAXTHRESHOLD = 0F;

   private int m_type;
   private float m_threshold;

   public Distortion()
   {
      m_type = HARDCLIP;
      m_threshold = -15F;
   }

   public int getType()
   {
```

```java
        return m_type;
    }

    public void setType(int type)
    {
        m_type = type;
    }

    public float getThreshold()
    {
        return m_threshold;
    }

    public void setThreshold(float threshold)
    {
        m_threshold = threshold;
    }

    public boolean allowsDryWetMix()
    {
        return true;
    }

    public boolean allowsSideChaining()
    {
        return false;
    }

    public boolean readObject(ReadInterface ar)
    {
        try
        {
            m_type = ar.readInt();
            m_threshold = ar.readFloat();
        }
        catch (IOException e)
        {
            return false;
        }

        return true;
    }

    public boolean writeObject(WriteInterface ar)
    {
        try
        {
            ar.writeInt(m_type);
```

```
            ar.writeFloat(m_threshold);
      }
      catch (IOException e)
      {
         return false;
      }

      return true;
   }

   public void setEqual(EffectInterface effect)
   {
      if (this == effect)
         return;
      if (effect.getClass() != this.getClass())
         return;
      Distortion e = (Distortion) effect;
      m_type = e.m_type;
      m_threshold = e.m_threshold;
   }

   public void startPlay()
   {
   }

   public void stopPlay()
   {
   }

   public boolean hasData()
   {
      return false;
   }

   public void setLanguage(String language)
   {
   }

   public void apply(byte [] dry, byte [] wet, byte [] control,
      AudioFormat format, double time)
   {
      // Initialize variables.  blockAlign is the size of the block align in the
      // audio (the number of bytes for one sample from all channels).  The
      // computation here is applicable to any sampling resolution or rate. sFrom
      // is the value of one sample in the incoming audio data.  sTo is the value
      // of one sample of the audio data produced by the effect.  sign is the
      // sign of the sample sFrom
```

```java
int blockAlign = (format.getChannels() * format.getSampleSizeInBits()) / 8;
int channels = format.getChannels();
float sFrom = 0;
float sTo = 0;
int sign = 1;

// Convert the threshold from decibels to a multiple.  For example, a gain
// of about 6 decibels will be translated to a multiple of approximately 2
float threshold = (float) Math.pow(10, m_threshold / 20);

// Loop through the samples in the signal.  In a mono 16-bit signal,
// blockAlign is 2 bytes
for(int i = 0; i < dry.length; i += blockAlign)
{
   // Compute the value and sign of each incoming sample
   sFrom = (float) ((short) (((dry[i + 1] & 0xff) << 8) + (dry[i] & 0xff)))
     / Short.MAX_VALUE;
   sign = (sFrom >= 0) ? 1 : -1;

   // Compute the value of output at the sample.  It is better if this
   // check encompasses and repeats the 'for' loop, so that the check is
   // not performed at every sample, but this version is easier to read
   sTo = 0;
   if (m_type == HARDCLIP)
   {
      // Cut the peaks of the signal at the threshold
      sTo = sign * Math.min(Math.abs(sFrom), threshold);
   }
   else if (m_type == SOFTCLIP)
   {
      // Apply the cubic soft clipper
      if (Math.abs(sFrom) <= threshold)
         sTo = sFrom - (sFrom * sFrom * sFrom / 3);
      else
         sTo = sign * (threshold - (threshold * threshold * threshold
            / 3));
   }
   else if (m_type == NOCLIP)
   {
      // Condense the peaks above the threshold
      if (Math.abs(sFrom) <= threshold)
         sTo = sFrom;
      else
         sTo = sign * (threshold + (float) Math.atan((Math.abs(sFrom) -
            threshold) / threshold) * threshold);
   }

   // Ensure that the computed output is within the limits of the bit
```

```
        // resolution
        sTo = Math.min(Math.max(sTo * Short.MAX_VALUE, Short.MIN_VALUE),
            Short.MAX_VALUE);

        // Store the computed output sample in the output buffer
        wet[i] = (byte) ((int) sTo & 0xff);
        wet[i + 1] = (byte) ((int) sTo >>> 8 & 0xff);
      }
    }
}
```

This class implements **EffectInterface**. With **EffectInterface**, Orinj can use many different effects without knowing exactly what they are.

String [] TYPES – These strings describe the types of distortion implemented here.

int HARDCLIP – This is a flag to use hard clip distortion.

int SOFTCLIP – This is a flag to use soft clip distortion.

int NOCLIP – This is a flag to use no clip distortion.

float MINTHRESHOLD – This is the minimum threshold in decibels that the user can choose, above which signal peaks will be clipped or condensed. The actual threshold chosen by the user is **m_threshold** below. **MINTHREHSHOLD** is used only to set up the graphical user interface and when checking user input for errors.

float MAXTHRESHOLD – This is the maximum threshold in decibels that the user can choose, above which signal peaks will be clipped or condensed.

int m_type – This is the type of distortion chosen by the user. It can be **HARDCLIP**, **SOFTCLIP**, or **NOCLIP**.

float m_threshold – This is the threshold, above which signal peaks will be clipped or condensed. It should be between **MINTHRESHOLD** and **MAXTHRESHOLD**.

Distortion() – This is the default and only constructor.

int getType() – This function returns the type of distortion **m_type**.

void setType(int type) – This function sets the type of distortion **m_type** to **type**.

float getThreshold() – This function returns the threshold **m_threshold**, above which peaks will be clipped or condensed.

void setThreshold(float threshold) – This function sets the threshold **m_threshold**, above which peaks will be clipped or condensed, to **threshold**.

boolean allowsDryWetMix() – This function is required by **EffectInterface**. It simply returns **true**, as separating the dry and wet signal in a simple delay is easy and doing so allows Orinj to mix those two signals at different levels if the user chooses to do so. Of course, the **apply** function below must be implemented accordingly. In most cases, Orinj users will probably mix the wet signal at its full amplitude and zero out the dry signal. Nonetheless, allowing a dry wet mix means that the user can change the combination of the two signals and make the distortion less audible.

boolean allowsSideChaining() – This function is required by **EffectInterface**. This effect does not use information from another signal and therefore does not allow side chaining. This function simply returns **false**.

void startPlay() – This function is required by **EffectInterface**. This function must implement actions that should be taken at the beginning of playing. There is nothing that needs to be done in this class.

void stopPlay() – This function is required by **EffectInterface**. Most Orinj DSP effects will not make use of this function.

boolean hasData() – This function is required by **EffectInterface**. In this class, this function should simply return **false**. This function is discussed further in other effects below.

void setLanguage(String language) – This function is required by **EffectInterface**. This function allows Orinj to set the language for effect controls and tooltips to the language selected by the user in the Orinj preferences. **language** is the ISO 639-2 three-letter code for the chosen language. Effects do not have to implement this function. It is not implemented above.

boolean writeObject(WriteInterace ar) – This function is required by **EffectInterface**. This function stores the effect member data for serialization and for undo. It returns **true** if there are no errors and **false** if there are errors. As above, **ar** can be used approximately as an object of class **java.io.ObjectOutputStream**.

boolean readObject(ReadInterace ar) – This function is required by **EffectInterface**. This function reads the effect member data from serialization and for undo. It returns **true**, if there are no errors and **false**, if there are errors. **ar** can be used approximately as an object of class **java.io.ObjectInputStream**.

void setEqual(EffectInterace effect) – This function is required by **EffectInterface**. This function sets the member data of the current effect equal to the member data of **effect**. This function is not used in the current version of Orinj, but may be used in future versions.

void apply(byte [] dry, byte [] wet, byte [] control, AudioFormat format, double time) – This function is required by **EffectInterface**. It applies the effect to the audio data contained in **dry**. To allow Orinj to mix the dry and wet audio data, **dry** is left unchanged and the delayed and decayed repetition of the signal is placed in **wet**. **dry**, **wet**, and **control** are of the same length and with audio data in the format specified by **format**. This effect does not allow side chaining and **control** is therefore not used. **dry** does not contain all audio data that are sent to the effect. Since audio data can be large, they are sent to the effect in pieces.

As with all effects throughout this book, this implementation only works for the sampling resolution of 16 bits. Other sampling resolutions are briefly discussed below.

While this implementation is appropriate for both mono (single channel) and stereo (two channel) signals, other effects must work with each channel separately. Working with mono and stereo signals is further described below.

This effect allows dry / wet mix: adjustments in the relative levels of the dry and wet signal. Orinj will take both the dry and wet signals and will mix those according to the dry and wet levels specified by the user. If an effect does not allow dry / wet mix, Orinj will take only the wet data and ignore the dry data in further processing.

This effect does not allow side chaining. A side chained effect is one that uses other sound data to control the sound data that are changed. A side chained compressor, for example, changes the dynamics of one track based on the dynamics of a second track. The audio data contained in the second track would be sent to the effect in the parameter control. In an effect that does not allow side chaining, control may be null and should not be used.

4.2. Using audio buffers

Take a two-and-a-half-minute song, recorded in two channels with the sampling rate 44100 Hz and sampling resolution 16 bits (2 bytes per sample per channel). The ready mixed song will contain 2.5 * 60 * 44100 * 2 * 2 = 26,460,000 bytes ≈ 25 Mb. Now suppose that you are recording this song in at least four tracks. The audio data for this short and simple recording could easily exceed 100 Mb. Actual recording sessions may be even larger.

Buffers allow audio software applications to keep audio data on disk, rather than in memory, and to only access and process portions of these data. Of course, it does not make sense to access and process the data byte by byte. This would require too many disk access operations, which would be slow. Instead, data are read in buffers that are larger than a single byte, but smaller than the whole song.

Using audio buffers in the implementation of this distortion is easy, as the distortion is computed from information contained only in the current buffers. This is not so with other effects. All remaining effects in this book must use not only the current buffers, but also past incoming audio buffers. They must therefore store and manage past audio buffers. Examples of how this can be done are presented in the chapters that follow.

4.3. Other sampling rates and resolutions

There is no need for changes in the code to handle other sampling rates. Adjustments are needed for other sampling resolutions.

Note the following code snippets. These convert byte data into audio samples and audio samples into bytes. In the distortion effect, there is additional scaling by **Short.MAX_VALUE** to ensure that the effect works with floating-point numbers between -1 and 1, but the important portions of the code are here below.

Code 12. Bytes to 16-bit samples and 16-bit samples to bytes

```
sFrom = (short) (((bufferFrom[curByte + 1] & 0xff) << 8)
    + (bufferFrom[curByte] & 0xff));
...
wet[i] = (byte)(sTo & 0xff);
wet[i + 1] = (byte)(sTo >>> 8 & 0xff);
```

Each sample in a 16-bit sampling resolution is stored in $16 / 8 = 2$ bytes. Since WAVE files are little endian, the least significant bit is first and can be used as it is. The second byte is the most significant byte and must be up-shifted by 8 bits. The sum of the two is the audio sample. The bit masks (0xff) are there to remove the sign – otherwise each byte will be treated as a signed byte – and the typecasting to short introduces the sign into the two-byte value. The opposite steps are taken when converting the value **sTo** to bytes.

Note that some (probably most) audio applications convert byte data to floating-point data and, internally, work only with floating-point buffers. In such an application, the apply function may have floating-point buffer arguments. This may be beneficial, as it would mean that the code does not have to be adjusted to handle other sampling resolutions. The code above does need changes.

The corresponding code for 8-bit (unsigned) audio data is as follows. The use of 128 is here, since this would be the middle of a signal oscillating between 0 and 256 in unsigned data represented by one byte samples. In addition, 128 represents silence, unlike with 16, 24, and 32 bits, where silence is zero. Note also that the limits applied to **sTo** in 8-bit computations are **Byte.MIN_VALUE** and **Byte.MAX_VALUE**, rather than **Short.MIN_VALUE** and **Short.MAX_VALUE**.

Code 13. Bytes to 8-bit samples and 8-bit samples to bytes

```
sFrom = (bufferFrom[curByte] & 0xff) -128;
...
wet[i] = (byte)((sTo + 128) & 0xff);
```

The corresponding code for 24-bit audio data is as follows. The limits applied to 24-bit data are ± 8388607. The upshifting by additional 8 bits than what might be expected and the downshifting by eight bits is to preserve the sign (i.e., up to a four-byte integer before downshifting to a three-byte integer).

Code 14. Bytes to 24-bit samples and 24-bit samples to bytes

```
sFrom = ((bufferFrom[curByte + 2] & 0xff) << 24)
    + ((bufferFrom[curByte + 1] & 0xff) << 16)
    + ((bufferFrom[curByte] & 0xff) << 8) >> 8;
...
wet[i] = (byte)(sTo & 0xff);
wet[i + 1] = (byte)(sTo >>> 8 & 0xff);
```

```
wet[i + 2] = (byte)(sTo >>> 16 & 0xff);
```

When working with 32-bit data, Orinj expects floating-point numbers. The byte-to-sample and back conversions are as follows (**fFrom** and **fTo** are floating-point numbers and **iTo** is an integer). The maximum and minimum values for the samples are ±1.

Code 15. Bytes to 32-bit samples and 32-bit samples to bytes

```
fFrom = Float.intBitsToFloat(
   ((bufferFrom[curByte + 3] & 0xff) << 24)
   + ((bufferFrom[curByte + 2] & 0xff) << 16)
   + ((bufferFrom[curByte + 1] & 0xff) << 8)
   + (bufferFrom[curByte] & 0xff));
...
iTo = Float.floatToIntBits(fTo);
wet[i] = (byte)(iTo & 0xff);
wet[i + 1] = (byte)(iTo >>> 8 & 0xff);
wet[i + 2] = (byte)(iTo >>> 16 & 0xff);
wet[i + 3] = (byte)(iTo >>> 24 & 0xff);
```

> Subsequent effects are only implemented on 16-bit data and we do not discuss 8-, 24- or 32-bit data below. These conversions from byte arrays to integer or floating-point data can be used in each of the effects below, to create the appropriate bit resolution implementation.

4.4. Error checking

To simplify the code above, there is no checking for errors. Checks could be added at least to functions that set the values of member data of the distortion. For example:

Code 16. Example error checking in the simple delay

```
public void setThreshold(float threshold)
{
   if (threshold < MINTHRESHOLD || threshold > MAXTHRESHOLD)
   {
      // Show some error if needed
   }
   threshold = Math.min(MAXTHRESHOLD, Math.max(MINTHRESHOLD, threshold));
   m_threshold = threshold;
}
```

> While it is beneficial to implement such checks in effects, for brevity, we do not discuss these types of error checks below.

4.5. Implementation of the distortion graphical user interface

The following is an implementation of the graphical user interface for the distortion in the Orinj framework.

Code 17. Distortion graphical user interface

```
package mycompany.com;

import java.awt.*;
import java.awt.event.*;
import java.util.Hashtable;
import javax.swing.*;
import javax.swing.event.*;
import javax.sound.sampled.*;
import recordingblogs.com.oreffect.*;

public class DistortionPanel extends JPanel implements EffectPanelInterface,
    ActionListener, ChangeListener, FocusListener
{
    private JComboBox<String> m_typeCombo;
    private JTextField m_thresholdField;
    private JSlider m_thresholdSlider;

    private Distortion m_distortion;

    public class SliderLabel extends JLabel
    {
        public SliderLabel(String string)
        {
            super(string);
            setFont(EffectFont.SMALLFONT);
        }
    }

    private class FocusPolicy extends
        ContainerOrderFocusTraversalPolicy
    {
        public Component getDefaultComponent(Container c)
        {
            return m_typeCombo;
        }

        public Component getLastComponent(Container c)
        {
            return m_thresholdSlider;
        }

        public Component getFirstComponent(Container c)
```

```
   {
      return m_typeCombo;
   }

   public Component getComponentBefore(Container c, Component a)
   {
      if (a == m_thresholdSlider)
         return m_thresholdField;
      if (a == m_thresholdField)
         return m_typeCombo;
      if (a == m_typeCombo)
         return m_thresholdSlider;
      return m_typeCombo;
   }

   public Component getComponentAfter(Container c, Component a)
   {
      if (a == m_typeCombo)
         return m_thresholdField;
      if (a == m_thresholdField)
         return m_thresholdSlider;
      if (a == m_thresholdSlider)
         return m_typeCombo;
      return m_typeCombo;
   }
}

public DistortionPanel(Distortion distortion, AudioFormat format)
{
   GridBagLayout layout = new GridBagLayout();
   GridBagConstraints constraints = new GridBagConstraints();
   setLayout(layout);

   JLabel typeLabel = new JLabel("Type:");
   typeLabel.setFont(EffectFont.LARGEFONT);
   constraints.gridx = 0;
   constraints.gridy = 0;
   constraints.weightx = 0;
   constraints.weighty = 1;
   constraints.anchor = GridBagConstraints.NORTHWEST;
   constraints.fill = GridBagConstraints.BOTH;
   layout.setConstraints(typeLabel, constraints);
   add(typeLabel);

   m_typeCombo = new JComboBox<String>(Distortion.TYPES);
   m_typeCombo.addActionListener(this);
   m_typeCombo.addFocusListener(this);
```

```java
m_typeCombo.setToolTipText("Set the type of distortion");
constraints.gridx = 2;
constraints.weightx = 1;
constraints.insets.left = 5;
layout.setConstraints(m_typeCombo, constraints);
add(m_typeCombo);

JLabel thresholdLabel = new JLabel("Threshold (dB):");
thresholdLabel.setFont(EffectFont.LARGEFONT);
constraints.gridx = 0;
constraints.gridy = 1;
constraints.weightx = 0;
constraints.insets.top = 5;
constraints.anchor = GridBagConstraints.NORTHWEST;
constraints.fill = GridBagConstraints.HORIZONTAL;
layout.setConstraints(thresholdLabel, constraints);
add(thresholdLabel);

m_thresholdField = new JTextField(4);
m_thresholdField.setToolTipText("Set the threshold");
m_thresholdField.addActionListener(this);
m_thresholdField.addFocusListener(this);
constraints.gridx = 1;
constraints.insets.left = 5;
layout.setConstraints(m_thresholdField, constraints);
add(m_thresholdField);

Hashtable<Integer, SliderLabel> numbersThreshold =
    new Hashtable<Integer, SliderLabel>();
numbersThreshold.put(new Integer(-50), new SliderLabel("-50"));
numbersThreshold.put(new Integer(-40), new SliderLabel("-40"));
numbersThreshold.put(new Integer(-30), new SliderLabel("-30"));
numbersThreshold.put(new Integer(-20), new SliderLabel("-20"));
numbersThreshold.put(new Integer(-10), new SliderLabel("-10"));
numbersThreshold.put(new Integer(0), new SliderLabel("0"));

m_thresholdSlider = new JSlider(JSlider.HORIZONTAL,
    (int) Distortion.MINTHRESHOLD, (int) Distortion.MAXTHRESHOLD, 0);
m_thresholdSlider.setMajorTickSpacing(10);
m_thresholdSlider.setMinorTickSpacing(5);
m_thresholdSlider.setPaintTicks(true);
m_thresholdSlider.setPaintLabels(true);
m_thresholdSlider.setFont(EffectFont.SMALLFONT);
m_thresholdSlider.setLabelTable(numbersThreshold);
m_thresholdSlider.setToolTipText("Set the threshold");
m_thresholdSlider.addChangeListener(this);
m_thresholdSlider.addFocusListener(this);
constraints.gridx = 2;
```

```java
      constraints.weightx = 1;
      layout.setConstraints(m_thresholdSlider, constraints);
      add(m_thresholdSlider);

      m_distortion = distortion;

      setFocusTraversalPolicy(new FocusPolicy());
      updateData();
   }

   public void focusGained(FocusEvent e)
   {
   }

   public void focusLost(FocusEvent e)
   {
      if (e.getSource() == m_thresholdField)
      {
         m_distortion.setThreshold(Float.parseFloat(m_thresholdField.getText()));
         updateData();
      }
      else if(e.getSource() == m_thresholdSlider)
      {
         m_distortion.setThreshold((float) m_thresholdSlider.getValue());
      }
      else if (e.getSource() == m_typeCombo)
      {
         m_distortion.setType(m_typeCombo.getSelectedIndex());
      }
   }

   public void stateChanged(ChangeEvent e)
   {
      if(e.getSource() == m_thresholdSlider)
      {
         m_distortion.setThreshold((float) m_thresholdSlider.getValue());
         updateData();
      }
   }

   public void actionPerformed(ActionEvent e)
   {
      if(e.getSource() == m_thresholdField)
      {
         m_distortion.setThreshold(Float.parseFloat(m_thresholdField.getText()));
         updateData();
      }
      else if(e.getSource() == m_typeCombo)
```

```
        {
            m_distortion.setType(m_typeCombo.getSelectedIndex(), false);
        }
    }

    public void updateData()
    {
        m_typeCombo.setSelectedIndex(m_distortion.getType());
        m_thresholdSlider.setValue((int) m_distortion.getThreshold());
        m_thresholdField.setText(Float.toString(m_distortion.getThreshold()));
    }
}
```

The member data and functions of this class are as follows.

JComboBox<String> m_typeCombo – This combo box allows the user to select the type of distortion – hard clip, soft clip, or no clip.

JTextField m_thresholdField – This field allows the user to adjust the threshold of the distortion.

JSlider m_thresholdSlider – This slider also allows the user to adjust the threshold of the distortion. Using a slider and a text field is a choice. The slider is easier to use, but not as precise. In this graphical user interface, as one of these two controls changes, the other one is adjusted accordingly.

Distortion m_distortion – This is the distortion effect.

class SliderLabel – This is a simple class to allow the threshold slider to use **EffectFont** fonts for its labels.

class FocusPolicy – This is the focus order of the user interface. Defining the focus policy is a good idea either way, but the focus policy of this user interface is used by Orinj in the dialog that encompasses this interface when the effect is created by the user.

DistortionPanel(Distortion distortion, AudioFormat format) – This is the default and only constructor. Note the call to **updateData** at the end of the constructor, which sets the controls in this interface to the values of the effect controls.

void focusLost(FocusEvent e) – This function captures the focus leaving and control and sets the corresponding control in the effect to the value of the interface control.

void stateChanged(ChangeEvent e) – This function captures changes in the threshold slider.

void actionPerformed(ActionEvent e) – This function captures changes in the combo box and pressing ENTER in the text field.

void updateData() – This function sets the interface controls to the values of the corresponding controls in the effect.

4.6. Error checking in the effect panel interface

The user can enter erroneous data in the text fields, such as "-4t" instead of "-45" for the threshold or "45" where the threshold can only be negative. Error checking can be implemented in **focusLost** as follows (the following is only a portion).

Code 18. Error checks in the graphical user interface – focusLost

```java
public void focusLost(FocusEvent e)
{
    if (e.getSource() == m_thresholdField)
    {
        try
        {
            float value = Float.parseFloat(m_thresholdField.getText());
            if (value < Distortion.MINTHRESHOLD || value > Distortion.MAXTHRESHOLD)
            {
                JOptionPane.showMessageDialog(this, "Please enter a number between "
                    + Distortion.MINTHRESHOLD + " and " + Distortion.MAXTHRESHOLD,
                    "Error", JOptionPane.ERROR_MESSAGE);
                updateData();
                m_thresholdField.grabFocus();
                return;
            }
            m_distortion.setThreshold(value);
            updateData();
        }
        catch (NumberFormatException ex)
        {
            JOptionPane.showMessageDialog(this, "Please enter a valid number",
                "Error", JOptionPane.ERROR_MESSAGE);
            updateData();
            m_thresholdField.grabFocus();
        }
    }
    ...
}
```

This code will show the first error if the user enters a number outside of the minimum and maximum allowed threshold values. The code will show the second error if the user enters a non-number. Without this error checking, the effect – specifically the conversion of the text field text to a floating-point value – may throw and uncaught exception.

Checking for errors in text fields should also be performed in **actionPerformed**. The following is an example. Note that on errors **actionPerformed** simply changes the current control focus to another control, which means that the error is caught and a message is displayed by **focusLost**.

Code 19. Error checks in the graphical user interface – actionPerformed

```
public void actionPerformed(ActionEvent e)
{
   if(e.getSource() == m_thresholdField)
   {
      try
      {
         float value = Float.parseFloat(m_thresholdField.getText());
         if (value < Distortion.MINTHRESHOLD || value > Distortion.MAXTHRESHOLD)
         {
            m_thresholdSlider.grabFocus();
            return;
         }
         m_distortion.setThreshold(value, false);
         updateData();
      }
      catch (NumberFormatException ex)
      {
         m_thresholdSlider.grabFocus();
      }
   }
   ...
}
```

> The code samples in the rest of the book do not include such error checks. This is done for brevity. These error checks are needed. They may be implemented as they are implemented here or differently.

4.7. Undo

To implement undo in the Orinj effect framework, when setting **m_threshold** and **m_type**, the effect class **Distortion** can fire an undo event as in the following example.

Code 20. Undo in the distortion when setting the threshold

```
public void setThreshold(float threshold, boolean storeUndo)
{
   if (threshold != m_threshold && storeUndo)
      fireEffectUndoEvent(new EffectUndoEvent(this,
         "Undo Distortion Threshold"));
   m_threshold = threshold;
}
```

This is possible, because **Distortion** extends **Undo**.

Note the addition of **storeUndo** in the function arguments. **storeUndo** is useful if we do not want to fire an undo event every time the threshold changes. In Orinj, for example, effect member data change with every movement of a corresponding control slider. This allows the user to hear changes in the effect as the slider moves during playback. However, an undo event is created only when the focus moves away from the slider and not on every movement of the slider (i.e., when the user moves to a different control).

In other words, **stateChanged** in the implementation of **DistortionPanel** calls **setThreshold** with **storeUndo** equal to **false**, but **focusLost** calls **setThreshold** with **storeUndo** equal to **true**. Note also that without **storeUndo**, there is no need to implement **focusLost** on the threshold slider as the threshold is set to its appropriate value by **stateChanged**.

> Implementing undo is not required. It is the programmer's choice. The code samples in the sections that follow do not implement undo.

4.8. Implementation in effect.xml

The following is an example of the effect.xml file that should be included in the effect package. More effects can be added in the same XML file. The two languages are also examples. More (or less) languages can be used.

Code 21. effect.xml file for the distortion

```xml
<?xml version="1.0" encoding="UTF-8"?>
<effectpack xmlns:xsi="http://www.w3.org/2001/XMLSchema-instance"
xsi:noNamespaceSchemaLocation="http://www.recordingblogs.com/sa/rbdocs/xml
orinjeffect.xsd">
   <effect>
      <effectclass>mycompany.com.Distortion</effectclass>
      <dialogclass>mycompany.com.DistortionPanel</dialogclass>
      <title>My Distortion</title>
      <version>3.0</version>
      <type>dynamics</type>
      <translations>
         <translation>
            <codeISO639-2>eng</codeISO639-2>
            <translatedtitle>My Distortion</translatedtitle>
         </translation>
         <translation>
            <codeISO639-2>bul</codeISO639-2>
            <translatedtitle>Мой дисторшън</translatedtitle>
         </translation>
      </translations>
   </effect>
</effectpack>
```

4.9. Packaging

The following builds a JAR file for an effect pack that contains a single effect – the distortion discussed in this chapter.

This code assumes that the path to **javac** has been set and that the directory **class** exists.

Code 22. Packaging the distortion

```
javac "Delay\src\mycompany\com\Distortion.java" -d "class" -classpath
    "class;Orange\oreffect.jar"
javac "Delay\src\mycompany\com\DistortionPanel.java" -d "class" -classpath
    "class;Orange\oreffect.jar"
copy src\effect.xml class\effect.xml
jar cf "exampledistortion.jar" -C "class" .
```

The resulting JAR file should be placed in the orange/effects folder of the installation for Orinj. Orinj will recognize it and use it automatically upon startup.

4.10. Obfuscating

If the effect code is obfuscated, the two classes – in this case **Distortion** and **DistortionPanel** – should be preserved (that is, not renamed). In **Distortion**, the obfuscation should also preserve the constructor **Distortion** and the function **setLanguage**. In **DistortionPanel**, the obfuscation should preserve the constructor **DistortionPanel**. The constructors are necessary so that Orinj can instantiate objects of these two classes. **setLanguage** is similarly necessary so that Orinj can set the language of the effect when the language in Orinj changes.

Chapter 5. Testing the distortion

There are two ways to test effects. Effect package JAR files can be placed in the effects folder of the Orinj installation and tested through Orinj. This does not allow debugging. Effects can also be tested with the simplified code presented below. This code is available for download from RecordingBlogs.com.

The package that contains the code is called ExampleDelayTest. It was designed to test ExampleDelay – a simple delay effect also available for download – but can be used to test any effect created with the Orinj effect framework. This source code replicates closely the way Orinj searches for effect JAR files, extracts information about the effects included in these JAR files, and creates the effect and its graphical user interface. The code also allows the playing of a single wave file and applying one of the available effects to the playback.

The classes in ExampleDelayTest are as follows. These classes are further described in detail below.

AudioBuffer – An implementation of a buffer of audio data.

EffectDialog – A dialog that contains the controls of the effect to be tested.

EffectStore – A Java **Vector** containing all available effects.

EffectStoreItem – A part of **EffectStore** that contains information about a single effect.

ExampleDelayTest – The main class of the package. This class loads effects and allows the user to choose and play a wave file with one of the effects.

MainFrame – A class used to load and instantiate effects.

Mixer – A class responsible for obtaining audio data from the wave file, applying the effect, and producing playback.

SelectDialog – A dialog to allow the user to choose an effect for playback.

WaveFile – An implementation of a wave file, used to obtain audio data from the wave file.

WaveFileDialog – A dialog to allow the user to choose a wave file for playback.

5.1. AudioBuffer.java

This class is a simple array of bytes that is used as a buffer to store audio data. Audio data are read from the wave file and sent to the output audio device. Since an audio buffer is not always full (e.g., when reading audio data from the end of the wave file), this class also keeps track of the number of bytes that were read and should be used during playback.

Code 23. AudioBuffer.java

```
package recordingblogs.com;
```

```
public class AudioBuffer
{
    private byte m_buffer[];
    private int m_actualSize;

    public AudioBuffer()
    {
        m_buffer = null;
        m_actualSize = 0;
    }

    public int getSize()
    {
        return m_buffer.length;
    }

    public void setSize(int size)
    {
        m_actualSize = size;
        m_buffer = new byte[size];
    }

    public byte[] getBuffer()
    {
        return m_buffer;
    }

    public int getActualSize()
    {
        return m_actualSize;
    }

    public void setActualSize(int size)
    {
        m_actualSize = size;
    }
}
```

The following are the member data and functions of this class.

byte [] m_buffer – This array of bytes contains the actual audio data.

int m_actualSize – If **m_buffer** contains less audio data than what its size suggests, **m_actualSize** is set accordingly. This may happen, for example, at the end of a wave file.

AudioBuffer() – This is the default and only constructor. In this constructor, the buffer is **null**. The buffer is created with the function **setSize** below, the call to which is external to this class. In Orinj, this call occurs when the user changes buffers sizes through the Orinj preferences.

int getSize() – This function returns the size of the buffer in bytes. This is the total size of the byte array, independent of how much of it contains audio data.

void setSize(int size) – This function sets the size of the buffer in bytes. The only concern is that the size is divisible by the block align of the audio data (the number of bytes in a single sample of audio for all channels, as discussed in chapter 2).

byte [] getBuffer() – This function provides access to the actual buffer.

int getActualSize() – This function returns the size of audio data in the buffer. While the number of bytes of audio data will typically be the same as the size of the byte array, it could be smaller if the reading of audio has reached the end of the wave file.

void setActualSize(int size) – This function sets the number of bytes of audio data contained in the buffer. This function is used typically only when audio data are read and the size of information read happens to be smaller than the buffer.

5.2. EffectDialog.java

This is an extension of **JDialog** that encompasses the graphical user interface for the effect. This class serves two purposes. First, the dialog shows up during playback, so that you can change the controls of the effect and test the effect. Second, in this application, closing the dialog causes playback to stop, so that you do not have to always listen to the whole wave file.

The controls for each effect are contained in a **JPanel** (e.g., **DistortionPanel** above). This dialog simply adds the panel and a Close button.

Note the import of the package **recordingblogs.com.oreffect**. oreffect.jar is needed for this application. oreffect.jar is included in the Orinj installation and can also be downloaded from RecordingBlogs.com. Its source code is described in chapter 3.

Code 24. EffectDialog.java

```java
package recordingblogs.com;

import java.awt.*;
import java.awt.event.*;
import javax.swing.*;
import recordingblogs.com.oreffect.*;

public class EffectDialog extends JDialog implements ActionListener
{
    private JButton m_closeButton;

    private Mixer m_mixer;
```

```java
public EffectDialog(EffectPanelInterface panel, Mixer mixer)
{
    super();

    m_mixer = mixer;

    Action escape = new AbstractAction()
    {
        public void actionPerformed(ActionEvent e)
        {
            m_mixer.stop();
            dispose();
        }
    };
    setTitle("Effect");
    setDefaultCloseOperation(DISPOSE_ON_CLOSE);

    GridBagLayout layout = new GridBagLayout();
    GridBagConstraints constraints = new GridBagConstraints();
    getContentPane().setLayout(layout);

    constraints.gridx = 0;
    constraints.gridy = 0;
    constraints.weightx = 1;
    constraints.weighty = 1;
    constraints.anchor = GridBagConstraints.NORTHWEST;
    constraints.fill = GridBagConstraints.BOTH;
    constraints.insets.left = 5;
    constraints.insets.top = 5;
    constraints.insets.bottom = 5;
    layout.setConstraints((JPanel) panel, constraints);
    getContentPane().add((JPanel) panel);

    m_closeButton = new JButton("Close");
    m_closeButton.addActionListener(this);
    m_closeButton.getInputMap(JComponent.WHEN_IN_FOCUSED_WINDOW).put(
        KeyStroke.getKeyStroke("ESCAPE"), "pressed ESCAPE");
    m_closeButton.getActionMap().put("pressed ESCAPE", escape);
    constraints.gridx = 1;
    constraints.weightx = 0;
    constraints.insets.right = 5;
    constraints.fill = GridBagConstraints.HORIZONTAL;
    layout.setConstraints(m_closeButton, constraints);
    getContentPane().add(m_closeButton);

    pack();
}
```

```
    public void actionPerformed(ActionEvent e)
    {
        if (e.getSource() == m_closeButton)
        {
            m_mixer.stop();
            dispose();
        }
    }
}
```

The following are the member data and functions in this class.

JButton m_closeButton – This is the Close button of the dialog. This button closes the dialog and stops playback.

Mixer m_mixer – The mixer applies the effect to audio data and plays the resulting audio.

EffectDialog(EffectPanelInterface panel, Mixer mixer) – This is the default and only constructor.

void actionPerformed(ActionEvent e) – This function handles the pressing of the close button.

5.3. EffectStore.java

This class is an array of the available effects. It replicates the way Orinj keeps track of available effects. As effects are read from the existing effect package JAR files, they are added to this array to be used later.

Code 25. EffectStore.java

```
package recordingblogs.com;

import java.util.Vector;

public class EffectStore
{
    public static Vector<EffectStoreItem> m_store;
}
```

5.4. EffectStoreItem.java

This class contains information about a single effect, including name, type, and the constructors for the effect and its graphical user interface. These items are created as effects are read from the existing effect package JAR files.

Code 26. EffectStoreItem.java

```
package recordingblogs.com;

import java.lang.reflect.Constructor;
```

```java
public class EffectStoreItem
{
    public final static String [] TYPE = {"delay", "reverb", "dynamics",
        "equalization", "filtering", "other"};

    private Constructor<?> m_effectConstructor;
    private Constructor<?> m_dialogConstructor;
    private String m_title;
    private String m_type;

    public EffectStoreItem(Constructor<?> effectConstructor,
        Constructor<?> dialogConstructor, String title, String type)
    {
        m_effectConstructor = effectConstructor;
        m_dialogConstructor = dialogConstructor;
        m_title = title;

        int index = 0;
        while(index < TYPE.length && (! type.equals(TYPE[index])))
            index++;
        if (index >= TYPE.length)
            type = TYPE[TYPE.length - 1];
        m_type = type;
    }

    public Constructor<?> getEffectConstructor()
    {
        return m_effectConstructor;
    }

    public Constructor<?> getDialogConstructor()
    {
        return m_dialogConstructor;
    }

    public String getTitle()
    {
        return m_title;
    }

    public String getType()
    {
        return m_type;
    }
}
```

This is simply a data storage class. As such, it consists mostly of data access functions.

`String [] TYPE` – These are the types of effects as described in the orinjeffect.xsd XML schema of chapter 3.

`Constructor<?> m_effectConstructor` – This is the constructor of the effect. This constructor is used to instantiate an object of the effect class when needed.

`Constructor<?> m_dialogConstructor` – This is the constructor of the effect graphical user interface. It is used to instantiate an object of the effect graphical user interface class when needed.

`String m_title` – This is the name of the effect as specified in effect.xml.

`String m_type` – This is the type of the effect as specified in effect.xml.

`EffectStoreItem(Constructor<?> effectConstructor, Constructor<?> dialogConstructor, String title, String type)` – This is the default and only constructor. In addition to setting all member data, this constructor also makes sure that the suggested effect type exists in the list of potential types. If not, the effect is assigned to "other".

`Constructor<?> getEffectConstructor()` – This function provides access to the constructor of the effect.

`Constructor<?> getDialogConstructor()` – This function provides access to the constructor of the effect graphical user interface.

`String getTitle()` – This function provides access to the effect title.

`String getType()` – This function provides access to the effect type.

5.5. ExampleDelayTest.java

This is the main function for the testing code. This function does the following.

- It loads all available effects by looking through the available effect packages.
- It shows all available effects to the user and allows the user to select an effect. The selected effect is the one that is applied during playback. If the user clicks on Cancel in the dialog for selecting an effect, the execution will stop.
- It allows the user to select a wave file to be played. If the user clicks on Cancel in the dialog for choosing a wave, the execution will stop.
- It opens the chosen wave and tests it to make sure that it is a valid wave file with the appropriate format. Note that this source code only works with wave files that are PCM wave files with compression code 1 and have the sampling rate 44100 Hz and the sampling resolution 16 bits per sample.
- It creates a mixer. The mixer applies the effect to the audio data in the wave file and plays the resulting audio.
- It creates the effect and the effect dialog and begins playback.

The following is the corresponding code.

Code 27. ExampleDelayTest.java

```java
package recordingblogs.com;

import java.io.*;
import java.util.*;
import java.lang.reflect.*;
import javax.swing.JOptionPane;
import recordingblogs.com.oreffect.*;

public class ExampleDelayTest
{
    public static void main(String[] args)
    {
        // Create a JFrame object and use it to load available effects.  There are
        // two reasons to use JFrame.  First, a JFrame loads effects in Orinj and
        // this code makes testing more reliable.  Second, an actual object must
        // be instantiated so that its class loader can be used
        MainFrame frame = new MainFrame();
        frame.loadEffects();

        // List all effects so the user can choose one to test
        Vector<String> listdata = new Vector<String>(0);
        for(int i = 0; i < EffectStore.m_store.size(); i++)
            listdata.add(EffectStore.m_store.get(i).getTitle());
        SelectDialog selectdlg = new SelectDialog(listdata);
        selectdlg.setModal(true);
        selectdlg.setVisible(true);
        if (! selectdlg.getOK())
            return;

        // Prompt the user to choose a wave for playback
        WaveFileDialog wavedlg = new WaveFileDialog();
        wavedlg.setModal(true);
        wavedlg.setVisible(true);
        if (! wavedlg.getOK())
            return;

        // Open the chosen WAVE file and check for the right format (e.g., a RIFF
        // file with a WAVE sub-chunk that contains a format and a data chunk).  No
        // audio data are processed here.  That happens below.  Since the effect
        // works only on wave files with signed PCM encoding and with 44100 Hz
        // sampling rate and 16-bit sampling resolution, the code also checks
        // for these
        WaveFile wave = null;
        try
        {
            wave = new WaveFile(wavedlg.getFileName().getText(),  "r");
```

```
      if (! wave.openRead())
      {
         System.out.println("Not the right wave file");
         wave.close();
         return;
      }
   }
   catch (IOException e)
   {
      System.out.println("Not the right wave file");
      return;
   }

   // Create a mixer.  This is a simplified mixer that can only play, but not
   // record, and can only play one wave and process one effect.  Send the
   // selected wave to the mixer
   Mixer mixer = new Mixer();
   mixer.setWave(wave);

   // Instantiate objects for the chosen effect and its graphical user
   // interface
   try
   {
      // The effect constructor does not have arguments
      EffectStoreItem item = EffectStore.m_store.get(selectdlg.getSelected());
      Object effect = item.getEffectConstructor().newInstance(
         (Object[]) null);

      // The graphical user interface JPanel requires two arguments - the
      // effect and the format of the wave
      Object [] arguments = new Object[2];
      arguments[0] = effect;
      arguments[1] = wave.getFormat();
      Object effectpanel = item.getDialogConstructor().newInstance(
         arguments);

      // Place the graphical user interface JPanel in a dialog that also
      // contains a Close button.  The Close button is used to close the
      // dialog and to stop playback
      EffectDialog effectdlg = new EffectDialog((EffectPanelInterface)
         effectpanel, mixer);
      effectdlg.setModal(false);
      effectdlg.setVisible(true);

      // Tell the mixer which effect to use
      mixer.setEffect((EffectInterface) effect);
   }
```

```
      catch (InvocationTargetException e)
      {
         JOptionPane.showInternalMessageDialog(null,
            "Cannot initialize the effect", "Error", JOptionPane.ERROR_MESSAGE);
         return;
      }
      catch (IllegalAccessException e)
      {
         JOptionPane.showInternalMessageDialog(null,
            "Cannot initialize the effect", "Error", JOptionPane.ERROR_MESSAGE);
         return;
      }
      catch (InstantiationException e)
      {
         JOptionPane.showInternalMessageDialog(null,
            "Cannot initialize the effect", "Error", JOptionPane.ERROR_MESSAGE);
         return;
      }

      // Start playback.  The mixer will obtain access to the output audio device
      // and will load it with some audio data
      if (! mixer.startPlay())
      {
         JOptionPane.showInternalMessageDialog(null,
            "Errors when attempting to play", "Error",
            JOptionPane.ERROR_MESSAGE);
         return;
      }

      // Continue playback.  The mixer will continue reading audio buffers and
      // sending them to the output audio device.  This is a separate thread so
      // that the user can continue working with the application and make
      // changes to the effect controls that will be heard during playback, as
      // well as press the Close button to stop playback if needed
      Thread thread = new Thread(mixer);
      thread.start();
   }
}
```

5.6. MainFrame.java

This class loads available effects. An extension of the class **JFrame** was chosen, as this is done similarly in Orinj and as we need the class loader of an instantiated object to load the constructors for the effect and its graphical user interface. This class does the following.

- It looks through the "effects" folder for JAR files.

- It unzips JAR files and looks for the file effect.xml in each JAR file. The effect.xml file should reside at the top of each effect package JAR file and should contain information on the effects in the specific JAR file.

- It loads all effects according to the information in the effect.xml files and places them in the effect store (see **EffectStore** and **EffectStoreItem** above).

Code 28. MainFrame.java

```java
package recordingblogs.com;

import java.io.*;
import java.lang.reflect.Constructor;
import java.net.*;
import java.util.Vector;
import java.util.jar.JarFile;
import java.util.zip.ZipEntry;
import java.util.zip.ZipInputStream;
import javax.swing.*;
import javax.xml.parsers.*;
import org.w3c.dom.Document;
import org.w3c.dom.NodeList;
import org.xml.sax.SAXException;

public class MainFrame extends JFrame
{
    public MainFrame()
    {
    }

    public static boolean isjar(File f)
    {
        if(f.isDirectory())
            return false;
        String extension = null;
        String s = f.getName();
        int i = s.lastIndexOf('.');
        if(i > 0 && i < s.length() - 1)
            extension = s.substring(i + 1).toLowerCase();
        if(extension != null)
            return extension.equals("jar");
        return false;
    }

    public void loadEffects()
    {
        // Use this to store information about all effects found
        EffectStore.m_store = new Vector<EffectStoreItem>(0);
```

```java
// Look for JAR files in the /effects folder.  If there are none, exit
File dir = new File("effects/");
File [] children = dir.listFiles();
if (children == null)
{
   JOptionPane.showMessageDialog(null, "Cannot find effect packasge JARs",
      "Error", JOptionPane.ERROR_MESSAGE);
   return;
}

// Loop through all files in the /effects folder.  Pick the JARs, look for
// the file effect.xml in the JARs, and load the effects
for (int i = 0; i < children.length; i++)
{
   // Check whether the file is a JAR file
   if (isjar(children[i]))
   {
      JarFile jf = null;
      ZipEntry zipEntry = null;
      ZipInputStream zipStream = null;
      FileInputStream fileStream = null;
      BufferedInputStream bufferStream = null;

      try
      {
         // Open the JAR file
         jf = new JarFile(children[i].getPath());

         // A JAR file is a zipped file.  Begin unzipping in search of
         // effect.xml
         fileStream = new FileInputStream(children[i].getPath());
         bufferStream = new BufferedInputStream(fileStream);
         zipStream = new ZipInputStream(bufferStream);
         zipEntry = zipStream.getNextEntry();

         // Loop through all items in the JAR file for effect.xml
         while ((zipStream != null) && (zipEntry != null)
           && (! zipEntry.getName().equals("effect.xml")))
         {
            if (zipStream.available() == 0)
               break;
            zipEntry = zipStream.getNextEntry();
         }

         // If we found effect.xml
         if (zipEntry != null && zipEntry.getName().equals("effect.xml"))
         {
```

```java
// Read the effect.xml file by parsing the information in the
// zip stream
DocumentBuilderFactory docBuilderFactory =
   DocumentBuilderFactory.newInstance();
DocumentBuilder docBuilder =
   docBuilderFactory.newDocumentBuilder();
Document doc = docBuilder.parse(zipStream);

// Get the contents of effect.xml according to the XML tags
NodeList effectclass = doc.getElementsByTagName("effectclass");
NodeList dialogclass = doc.getElementsByTagName("dialogclass");
NodeList title = doc.getElementsByTagName("title");
NodeList type = doc.getElementsByTagName("type");

// Just in case
if (effectclass.getLength() != dialogclass.getLength()
   || effectclass.getLength() != title.getLength()
   || effectclass.getLength() != type.getLength())
{
   JOptionPane.showInternalMessageDialog(null,
      "Mismatch when reading effects", "Error",
      JOptionPane.ERROR_MESSAGE);
   jf.close();
   return;
}

// Loop through the effects and load them
for(int j = 0; j < effectclass.getLength(); j++)
{
   // Load the effect class and the effect panel class
   String filePath = children[i].getPath();
   URI uri = new File(filePath).toURI();
   URL [] url = new URL[1];
   url[0] = uri.toURL();

   // Use the frame class loader
   URLClassLoader child = new URLClassLoader(url,
      this.getClass().getClassLoader());
   Class<?> classToLoad = Class.forName(
      effectclass.item(j).getTextContent(), true, child);
   Constructor<?> [] ce = classToLoad.getConstructors();
   classToLoad = Class.forName(
      dialogclass.item(j).getTextContent(), true, child);
   Constructor<?> [] cd = classToLoad.getConstructors();

   // Check that the size of ce and cd here is at least 1 and
   // that they are not null
```

```java
            if (ce == null || cd == null)
            {
                JOptionPane.showInternalMessageDialog(null,
                    "Cannot find effects", "Error",
                    JOptionPane.ERROR_MESSAGE);
                zipStream.close();
                bufferStream.close();
                fileStream.close();
                return;
            }
            if (ce.length < 1 || cd.length < 1)
            {
                JOptionPane.showInternalMessageDialog(null,
                    "Mismatch when reading effects", "Error",
                    JOptionPane.ERROR_MESSAGE);
                zipStream.close();
                bufferStream.close();
                fileStream.close();
                return;
            }

            // There may be other things we would like to do here, such
            // as choose the right language for the effect, if the
            // effect supports it.  For now, simply add the information
            // to the effect store
            EffectStore.m_store.add(new EffectStoreItem(ce[0], cd[0],
                title.item(j).getTextContent(),
                type.item(j).getTextContent()));
        }
    }

    // Close the zip stream
    if (zipStream != null)
        zipStream.close();
    if (bufferStream != null)
        bufferStream.close();
    if (fileStream != null)
        fileStream.close();
}
catch (IOException e)
{
    JOptionPane.showInternalMessageDialog(null,
        "Exception when loading JAR: " + children[i].getPath() + ": "
        + e, "Error", JOptionPane.ERROR_MESSAGE);
}
catch (SAXException e)
{
    JOptionPane.showInternalMessageDialog(null,
```

```
            "Exception when reading effect.xml in: " + jf.getName() + ": "
            + e, "Error", JOptionPane.ERROR_MESSAGE);
      }
      catch (ParserConfigurationException e)
      {
         JOptionPane.showInternalMessageDialog(null,
            "Exception when reading effect.xml in: " + jf.getName() + ": "
            + e, "Error", JOptionPane.ERROR_MESSAGE);
      }
      catch (ClassNotFoundException e)
      {
         JOptionPane.showInternalMessageDialog(null,
            "Exception when reading effect.xml in: " + jf.getName() + ": "
            + e, "Error", JOptionPane.ERROR_MESSAGE);
      }

      // Close the JAR file
      try
      {
         jf.close();
      }
      catch (IOException e)
      {
         JOptionPane.showInternalMessageDialog(null,
            "Exception when loading JAR: " + children[i].getPath() + ": "
            + e, "Error", JOptionPane.ERROR_MESSAGE);
         return;
      }
    }
   }
  }
}
```

The member functions of this class are as follows.

MainFrame() – This is the default and only constructor. The code does not actually invoke a **JFrame** and hence there is nothing to do in this constructor. The class loader of the frame is used, replicating the Orinj code, to load the effect pack JAR files.

boolean isjar(File f) – This function returns true if **f** is a JAR file and false otherwise.

void loadEffects() – This function loads the effects. It loops through all JAR files in the effects folder and, if these files can be unzipped and contain the appropriate effect.xml file, it loads the corresponding effects.

5.7. Mixer.java

This class applies the chosen effect to the chosen wave file and plays the resulting audio. A good amount of preparation for playback is done inside this class, including aligning the wave file pointer to the start of its data chunk, filling the audio buffers with silence, reading a few audio buffers from the wave file, and sending these buffers to the output audio device. This means that the output audio device's own buffer is filled with some amount of audio, which is played while the mixer reads and sends additional buffers.

The continuation of playback, after the initial preparation, is done in a separate thread. This is so to allow the user to change controls in the effect dialog (see **EffectDialog** above) or to press the Close button of the effect dialog and stop playback.

In principle, it is not necessary to send several buffers to the output audio device at the start of playback. If at least one buffer is sent and if the remaining code is executed sufficiently quickly, then the additional buffers will come from the separate playback thread while the first buffer is still playing, thus ensuring that there are no gaps in the playback.

Note that the mixer treats the effect as an object of class **EffectInterface**. This means that the mixer only has access to the functions of **EffectInterface** (in this case, the mixer only uses **apply** and **allowsDryWetMix**).

For effects that allow dry and wet mix, this mixer must mix the incoming dry buffer with the wet buffer from the effect (at which point, in Orinj, the user defined dry and wet mix are applied, even though there is no actual dry and wet mix here).

The mixing implemented in this code is only for 16-bit audio data.

Code 29. Mixer.java

```
package recordingblogs.com;

import javax.sound.sampled.*;
import recordingblogs.com.oreffect.EffectInterface;
import java.io.IOException;

public class Mixer implements Runnable
{
    private AudioBuffer [] m_bufferPlay;
    private AudioBuffer m_bufferWet;
    private WaveFile m_wave;
    private SourceDataLine m_line;
    private boolean m_playing;
    private EffectInterface m_effect;

    public Mixer()
    {
        // Choose a random size for the buffers.  The only concerns here are that
        // the buffer sizes are divisible by the block align and that the sizes of
```

```
    // the dry buffers and wet buffers are the same.  Of course, if the buffers
    // are very long, responses to user actions may be slow
    m_bufferPlay = new AudioBuffer [4];
    for(int i = 0; i < 4; i++)
    {
        m_bufferPlay[i] = new AudioBuffer();
        m_bufferPlay[i].setSize(16384);
    }
    m_bufferWet = new AudioBuffer();
    m_bufferWet.setSize(16384);
    m_wave = null;
    m_line = null;
    m_playing = false;
    m_effect = null;
}

public void setWave(WaveFile wave)
{
    m_wave = wave;
}

public void setEffect(EffectInterface effect)
{
    m_effect = effect;
}

public void stop()
{
    m_playing = false;
}

public boolean startPlay()
{
    // Line up the reading of the wave to the beginning of its data chunk
    m_wave.startPlay();

    // Fill the buffers with silence, just in case
    for(int i = 0; i < m_bufferPlay.length; i++)
    {
        for(int j = 0; j < m_bufferPlay[i].getSize(); j++)
            m_bufferPlay[i].getBuffer()[j] = 0;
    }

    // Read a few buffers from the wave
    for(int i = 0; i < m_bufferPlay.length; i++)
        m_wave.getData(m_bufferPlay[i]);
```

```
    // Open the output audio device
    try
    {
        m_line = (SourceDataLine)
            AudioSystem.getSourceDataLine(m_wave.getFormat());
        m_line.open();
        m_line.start();
    }
    catch (LineUnavailableException e)
    {
        return false;
    }

    // Feed the output audio device with the sound data already read
    for(int i = 0; i < m_bufferPlay.length; i++)
        m_line.write(m_bufferPlay[i].getBuffer(), 0,
            m_bufferPlay[i].getActualSize());

    // We are playing
    m_playing = true;

    return true;
}

public void run()
{
    // While we have not been told to stop
    while(m_playing)
    {
        // Read a buffer from the wave file
        m_wave.getData(m_bufferPlay[0]);

        // Check if this is the end of the wave file
        if (m_bufferPlay[0].getActualSize() < m_bufferPlay[0].getSize())
        {
            // Even if this is the end of the wave file, check if there is
            // residual audio in the effect
            if (! m_effect.hasData())
                m_playing = false;
        }

        // Apply the effect
        m_effect.apply(m_bufferPlay[0].getBuffer(), m_bufferWet.getBuffer(),
            null, m_wave.getFormat());

        // If the effect allows dry and wet mix, mix the dry and wet buffers
        // together
        if (m_effect.allowsDryWetMix())
```

```
        {
            float sFrom = 0, sTo = 0;
            int blockAlign = m_wave.getFormat().getSampleSizeInBits() / 8;
            for(int i = 0; i < m_bufferPlay[0].getActualSize(); i += blockAlign)
            {
                // Get a sample from the wet buffer
                sTo = (short)(((m_bufferWet.getBuffer()[i + 1] & 0xff) << 8)
                    + (m_bufferWet.getBuffer()[i] & 0xff));

                // Get a sample from the dry buffer
                sFrom = (short) (((m_bufferPlay[0].getBuffer()[i + 1] & 0xff)
                    << 8) + (m_bufferPlay[0].getBuffer()[i] & 0xff));

                // Add the two samples and make sure that they do not exceed
                // the limits of the sampling resolution
                sTo = (int)Math.min(Math.max(sFrom + sTo, Short.MIN_VALUE),
                    Short.MAX_VALUE);

                // Put the result in the output
                m_bufferWet.getBuffer()[i] = (byte) ((int) sTo & 0xff);
                m_bufferWet.getBuffer()[i + 1] = (byte)((int) sTo >>> 8 & 0xff);
            }
        }

        // Send the result to the output audio device. Do not use getActualSize
        // for the buffer here, as the actual size of the last buffer may be -1
        m_line.write(m_bufferWet.getBuffer(), 0, m_bufferPlay[0].getSize());
    }

    // Close the output audio device
    m_line.drain();
    m_line.stop();
    m_line.close();

    // Close the wave
    try
    {
        m_wave.close();
    }
    catch (IOException e)
    {
        System.out.println("Exception in Mixer::run: " + e);
    }
  }
}
```

The following are the member data and functions in this class.

AudioBuffer [] m_bufferPlay – This audio buffer contains incoming audio data.

AudioBuffer m_bufferWet – This buffer contains the data sent to the output sound device.

WaveFile m_wave – This is the wave file that is played.

SourceDataLine m_line – This is the playback sound device.

boolean m_playing – This variable is a way to communicate between threads whether there is playback.

EffectInterface m_effect – This is the effect that is applied.

Mixer() – This is the default and only constructor for the class. It prepares the audio buffers and initializes the remaining member data.

void setWave(WaveFile wave) – This function sets the wave that will be played to **wave**.

void setEffect(EffectInterface effect) – This function sets the effect that will be applied to **effect**.

void stop() – This function stops playback.

boolean startPlay() – This function starts playback. It prepares the wave and buffer for playback, opens the output sound device, and feeds the output sound devices with some data. This function returns **true** if playback started with no issues and **false** otherwise.

void run() – This is the implementation of playback. The use of the **Runnable** interface ensures that playback runs in a separate thread and the user can continue to work with the application (e.g., the user can press the Close button). During playback, the mixer reads additional audio data from the wave file, applies the effect to them, and sends the result to the output sound device. Playback stops when the audio data are exhausted (or earlier, if the user does so through the graphical user interface).

Playback works, because of the buffering and multithreading that happens in the output sound device. We are using a small buffer of only 16384 bytes. In one channel 44100 Hz audio, this buffer can store less than 0.2 seconds of audio. However, it takes only milliseconds to write (copy) this buffer to the **SourceDataLine m_line**. While **m_line** is busy playing this buffer from its own internal storage in a separate thread, the mixer has time to fill the buffer with new data, apply the effect, and send the result to **m_line** again.

The separate thread implemented with **Runnable** above should not be confused with the thread that **m_line** uses internally to play. The thread above is implemented simply to allow user input – to stop playback or to change the effect controls.

Playback and the reading of audio data do not have to be synchronized. If the storage internal to **m_line** is full, then **m_line.write** will wait, effectively blocking the mixer, until space is available. This means that we should not be concerned that the mixer may read and prepare data too fast. We should only be concerned if the mixer reads and prepares data too slowly. If the mixer is too slow, **m_line** could run out of information to play and produce a pause in the playback. The mixer of a larger application can be slow if there are several tracks of audio that should be mixed into one and multiple effects. Then, we can speed up the mixer, for example, with the user of larger buffers. Larger buffers reduce the number of times the mixer accesses the disk.

5.8. SelectDialog.java

This dialog simply lists the names of all available effects and lets the user choose one effect to be applied during playback.

Code 30. SelectDialog.java

```java
package recordingblogs.com;

import java.awt.*;
import java.awt.event.*;
import java.util.Vector;
import javax.swing.*;

public class SelectDialog extends JDialog implements ActionListener
{
    private JList<String> m_list;
    private JButton m_okButton;
    private JButton m_cancelButton;

    private boolean m_OK;
    private int m_selected;

    public SelectDialog(Vector<String> listData)
    {
        super();

        Action escape = new AbstractAction()
        {
            public void actionPerformed(ActionEvent e)
            {
                m_OK = false;
                dispose();
            }
        };

        setTitle("Select");
        setDefaultCloseOperation(DISPOSE_ON_CLOSE);

        GridBagLayout layout = new GridBagLayout();
        GridBagConstraints constraints = new GridBagConstraints();
        getContentPane().setLayout(layout);
```

```
JPanel buttonPanel = new JPanel();
GridBagLayout buttonLayout = new GridBagLayout();
GridBagConstraints buttonConstraints = new GridBagConstraints();
buttonPanel.setLayout(buttonLayout);
m_okButton = new JButton("OK");
m_okButton.addActionListener(this);
buttonConstraints.gridx = 0;
buttonConstraints.gridy = 0;
buttonConstraints.insets.right = 5;
buttonConstraints.anchor = GridBagConstraints.NORTH;
buttonConstraints.fill = GridBagConstraints.HORIZONTAL;
buttonLayout.setConstraints(m_okButton, buttonConstraints);
buttonPanel.add(m_okButton);

m_cancelButton = new JButton("Cancel");
m_cancelButton.addActionListener(this);
m_cancelButton.getInputMap(JComponent.WHEN_IN_FOCUSED_WINDOW).put(
    KeyStroke.getKeyStroke("ESCAPE"), "pressed ESCAPE");
m_cancelButton.getActionMap().put("pressed ESCAPE", escape);
buttonConstraints.gridy = 1;
buttonConstraints.insets.top = 5;
buttonLayout.setConstraints(m_cancelButton,
    buttonConstraints);
buttonPanel.add(m_cancelButton);

JPanel tempPanel = new JPanel();
buttonConstraints.weighty = 1;
buttonConstraints.gridy = 2;
buttonLayout.setConstraints(tempPanel, buttonConstraints);
buttonPanel.add(tempPanel);

m_list = new JList<String>();
m_list.setToolTipText("Select an item");
m_list.setSelectionMode(ListSelectionModel.SINGLE_SELECTION);
m_list.setListData(listData);
m_list.setBorder(BorderFactory.createLoweredBevelBorder());
m_list.setPrototypeCellValue("Some prototype value");

JScrollPane listScrollPane = new JScrollPane();
listScrollPane.setViewportView(m_list);
constraints.insets.left = 5;
constraints.insets.top = 5;
constraints.insets.bottom = 5;
constraints.gridx = 0;
constraints.gridy = 0;
constraints.weightx = 1;
constraints.weighty = 1;
```

71

```
            constraints.fill = GridBagConstraints.BOTH;
            layout.setConstraints(listScrollPane, constraints);
            getContentPane().add(listScrollPane);

            constraints.gridx = 1;
            constraints.weightx = 0;
            constraints.weighty = 0;
            constraints.anchor = GridBagConstraints.NORTH;
            constraints.fill = GridBagConstraints.VERTICAL;
            layout.setConstraints(buttonPanel, constraints);

            getContentPane().add(buttonPanel);

            m_OK = false;
            m_selected = 0;
            if(listData.size() <= 0)
               m_list.setEnabled(false);
            else
               m_list.setSelectedIndex(0);

            pack();
            m_list.ensureIndexIsVisible(m_list.getSelectedIndex());
        }

        public boolean getOK()
        {
            return m_OK;
        }

        public int getSelected()
        {
            return m_selected;
        }

        public void actionPerformed(ActionEvent e)
        {
            if(e.getSource() == m_okButton)
            {
               m_OK = true;
               m_selected = m_list.getSelectedIndex();
               dispose();
            }
            else if(e.getSource() == m_cancelButton)
            {
               m_OK = false;
               dispose();
            }
```

```
        }
}
```

5.9. WaveFile.java

The wave file is implemented as a **RandomAccessFile**. This is a good idea, as wave files consists of chunks, and some these chunks may be skipped. In this book, we use the format and data chunks of the wave file, but a wave file may contain other chunks, such as ones for markers, text information about the file, or other (see Appendix A).

The implementation below is sufficient for this book, but is not the best implementation. A wave file may contain a wave list chunk with alternating data and silence chunks. In this case, the data chunks are not at the top of the wave file, but are the sub-chunks of the wave list chunk. For such a file, the implementation below does not work. A better wave file implementation is a stack of chunks.

Code 31. WaveFile.java

```java
package recordingblogs.com;

import java.io.*;
import javax.sound.sampled.*;

public class WaveFile extends RandomAccessFile
{
    private long m_datapointer;
    private AudioFormat m_format;

    public WaveFile(String name, String rw) throws IOException
    {
        super(name, rw);
        m_datapointer = -1;
    }

    public AudioFormat getFormat()
    {
        return m_format;
    }

    public String readRIFFString(int size) throws IOException
    {
        String string = new String();
        for(int i = 0; i < size; i++)
        {
            char c = (char) readByte();
            string = string + c;
        }
        return string;
```

```
    }

    public int readRIFFUnsignedShort() throws IOException
    {
        byte b0 = readByte();
        byte b1 = readByte();
        return ((int) b0 + (int) (b1 << 4));
    }

    public long readRIFFUnsignedInt() throws IOException
    {
        long b0 = readUnsignedByte();
        long b1 = readUnsignedByte();
        long b2 = readUnsignedByte();
        long b3 = readUnsignedByte();
        return b0 + (b1 << 8) + (b2 << 16) + (b3 << 24);
    }

    public boolean openRead()
    {
        try
        {
            // Start from the beginning
            seek(0);

            // Check for a RIFF header
            String temp = readRIFFString(4);
            if(temp.compareTo("RIFF") != 0)
            {
                close();
                return false;
            }

            // Read the size of the RIFF chunk
            long size = readRIFFUnsignedInt();

            // Check the subtype of the RIFF chunk
            temp = readRIFFString(4);
            if(temp.compareTo("WAVE") != 0)
            {
                close();
                return false;
            }

            m_datapointer = -1;
            long fmtpointer = -1;
```

```
    // Look for a fmt chunk or a data chunk.  If the file does not have
    // both chunks, this code will throw an exception.  The exception is
    // caught below
    while(m_datapointer < 0 || fmtpointer < 0)
    {
        // Read the type of the chunk and check if this is a data or a format
        // chunk
        temp = readRIFFString(4);
        if(temp.compareTo("fmt ") == 0)
            fmtpointer = getFilePointer() + 4;
        if(temp.compareTo("data") == 0)
            m_datapointer = getFilePointer() + 4;
        size = readRIFFUnsignedInt();

        // Move to the next chunk
        seek(getFilePointer() + size);
    }

    // Read the format chunk
    seek(fmtpointer);
    short formatType = (short) readRIFFUnsignedShort();
    short channels = (short) readRIFFUnsignedShort();
    long sampleRate = readRIFFUnsignedInt();
    readRIFFUnsignedInt();    // reads bytes per second
    readRIFFUnsignedShort();  // reads bytes per sample
    short bitsPerChannel = (short) readRIFFUnsignedShort();

    // Check the format.  For this application, we want a PCM wave with the
    // sampling rate 44100 Hz and the 16-bit resolution
    if (formatType != 1 || sampleRate != 44100 || bitsPerChannel != 16)
    {
        close();
        return false;
    }

    // Alternatively, we can read the format chunk with the following
    // AudioSystem.getAudioFileFormat(new
    //     File(wavedlg.getFileName().getText()));

    // remember the format
    m_format = new AudioFormat(AudioFormat.Encoding.PCM_SIGNED, sampleRate,
        bitsPerChannel, channels, (bitsPerChannel / 8) * channels,
        sampleRate, false);
}
catch(IOException e)
{
    return false;
}
```

```
        return true;
    }

    // At the beginning of playback, line up to the beginning of the data chunk
    public boolean startPlay()
    {
        try
        {
            seek(m_datapointer);
        }
        catch (IOException e)
        {
            return false;
        }

        return true;
    }

    // Read a buffer of audio data
    public boolean getData(AudioBuffer buffer)
    {
        try
        {
            int actualsize = this.read(buffer.getBuffer(), 0, buffer.getSize());
            buffer.setActualSize(actualsize);
        }
        catch (IOException e)
        {
            return false;
        }

        return true;
    }
}
```

The following are the member data and functions of the class.

long m_datapointer – This is the position of the data chunk in the wave file (after the name of the chunk "data" and the four-byte size of the chunk). We keep this information so that we can quickly go to the audio data portion of the wave to begin playback.

AudioFormat m_format – This is the audio format of the wave. In many audio applications, quick access to the format is necessary so that the wave can be appropriately mixed. For example, one should know whether the wave contains one or two channels so that effects that use different parameters for the different channels can be applied.

String readRIFFString(int size) – This function reads a string from a random access file (the wave file). The names of RIFF file chunks are strings.

int readRIFFUnsignedShort() – This function reads an unsigned short integer from a random access file (the wave file). Unsigned shorts are contained in the format chunk of the wave file (e.g., compression code / format type, the number of channels).

long readRIFFUnsignedInt() – This function reads unsigned integers from a random access file (the wave file). The names of chunks are strings. Unsigned integers are used for the size of chunks and are also contained in the format chunk (e.g., sampling rate, etc.).

boolean openRead() – This function checks the format of the file (for the right headers, sub-headers, and chunks) and prepares the file for use (by aligning **m_datapointer**). Since the rest of the code only handles 16-bit PCM waves with the 44100 Hz sampling rate, this function also checks for the appropriate compression code, sampling resolution, and sampling rate.

boolean startPlay() – This function moves the file pointer to the beginning of the audio data.

boolean getData() – This function reads a buffer of audio data. This function is simple in this implementation, but could become a lot more complex if the application is to allow fast forwarding, rewinding, and looped playback.

5.10. WaveFileDialog.java

This dialog allows the user to choose a wave file for playback.

Code 32. WaveFileDialog.java

```java
package recordingblogs.com;

import java.awt.*;
import java.awt.event.*;
import java.io.File;
import javax.swing.*;

public class WaveFileDialog extends JDialog implements ActionListener
{
    private JTextField m_fileName;
    private JButton m_browseButton;
    private JButton m_okButton;
    private JButton m_cancelButton;
    private boolean m_ok;

    public WaveFileDialog()
    {
        super();
        setTitle("Wave File");

        // handle the user pressing escape
```

```java
Action escape = new AbstractAction()
{
   public void actionPerformed(ActionEvent e)
   {
      m_ok = false;
      dispose();
   }
};

m_ok = false;

GridBagLayout layout = new GridBagLayout();
GridBagConstraints constraints = new GridBagConstraints();
getContentPane().setLayout(layout);

JPanel titlesPane = new JPanel();
GridBagLayout titlesLayout = new GridBagLayout();
GridBagConstraints titlesConstraints = new GridBagConstraints();
titlesPane.setLayout(titlesLayout);

JLabel fileLabel = new JLabel("File: ");
titlesConstraints.gridx = 0;
titlesConstraints.gridy = 0;
titlesConstraints.weightx = 0;
titlesConstraints.weighty = 1;
titlesConstraints.anchor = GridBagConstraints.NORTHWEST;
titlesConstraints.fill = GridBagConstraints.HORIZONTAL;
titlesLayout.setConstraints(fileLabel, titlesConstraints);
titlesPane.add(fileLabel);

m_fileName = new JTextField(20);
m_fileName.setToolTipText("Enter a new file name");
titlesConstraints.gridx = 1;
titlesConstraints.weightx = 1;
titlesConstraints.insets.left = 5;
titlesLayout.setConstraints(m_fileName, titlesConstraints);
titlesPane.add(m_fileName);

m_browseButton = new JButton("Browse");
m_browseButton.addActionListener(this);
m_browseButton.setToolTipText("Click to browse for a specific file");
titlesConstraints.gridx = 2;
titlesConstraints.weightx = 0;
titlesLayout.setConstraints(m_browseButton, titlesConstraints);
titlesPane.add(m_browseButton);

constraints.gridx = 0;
```

```
    constraints.gridy = 0;
    constraints.weightx = 1;
    constraints.weighty = 0;
    constraints.anchor = GridBagConstraints.NORTHWEST;
    constraints.fill = GridBagConstraints.HORIZONTAL;
    constraints.insets.top = 5;
    constraints.insets.left = 5;
    layout.setConstraints(titlesPane, constraints);
    getContentPane().add(titlesPane);

    JPanel okPane = new JPanel();
    GridBagLayout okLayout = new GridBagLayout();
    GridBagConstraints okConstraints = new GridBagConstraints();
    okPane.setLayout(okLayout);

    m_okButton = new JButton("OK");
    m_okButton.addActionListener(this);
    okConstraints.gridx = 0;
    okConstraints.gridy = 0;
    okConstraints.weightx = 1;
    okConstraints.weighty = 1;
    okConstraints.fill = GridBagConstraints.HORIZONTAL;
    okConstraints.anchor = GridBagConstraints.NORTHWEST;
    okLayout.setConstraints(m_okButton, okConstraints);
    okPane.add(m_okButton);

    constraints.gridx = 1;
    constraints.weightx = 0;
    constraints.anchor = GridBagConstraints.NORTHWEST;
    constraints.fill = GridBagConstraints.HORIZONTAL;
    constraints.insets.right = 5;
    layout.setConstraints(okPane, constraints);
    getContentPane().add(okPane);

    JPanel cancelPane = new JPanel();
    GridBagLayout cancelLayout = new GridBagLayout();
    GridBagConstraints cancelConstraints = new GridBagConstraints();
    cancelPane.setLayout(cancelLayout);

    m_cancelButton = new JButton("Cancel");
    m_cancelButton.addActionListener(this);
    m_cancelButton.getInputMap(JComponent.WHEN_IN_FOCUSED_WINDOW).put(
        KeyStroke.getKeyStroke("ESCAPE"), "pressed ESCAPE");
    m_cancelButton.getActionMap().put("pressed ESCAPE", escape);
    cancelConstraints.gridx = 0;
    cancelConstraints.gridy = 0;
    cancelConstraints.weightx = 1;
    cancelConstraints.weighty = 0;
```

```java
        cancelConstraints.fill = GridBagConstraints.HORIZONTAL;
        cancelConstraints.anchor = GridBagConstraints.NORTHWEST;
        cancelLayout.setConstraints(m_cancelButton, cancelConstraints);
        cancelPane.add(m_cancelButton);

        JPanel emptyPane = new JPanel();
        cancelConstraints.gridy = 1;
        cancelConstraints.weighty = 1;
        cancelLayout.setConstraints(emptyPane, cancelConstraints);
        cancelPane.add(emptyPane);

        constraints.gridy = 1;
        constraints.weighty = 1;
        constraints.insets.bottom = 5;
        constraints.fill = GridBagConstraints.BOTH;
        layout.setConstraints(cancelPane, constraints);
        getContentPane().add(cancelPane);

        pack();
    }

    // Get whether the user pressed OK
    public boolean getOK()
    {
        return m_ok;
    }

    // Get the name of the file that the user chose (full name, including path)
    public JTextField getFileName()
    {
        return m_fileName;
    }

    // Capture the user pressing the three buttons
    public void actionPerformed(ActionEvent e)
    {
        if (e.getSource() == m_okButton)
        {
            m_ok = true;
            dispose();
        }
        else if(e.getSource() == m_cancelButton)
        {
            m_ok = false;
            dispose();
        }
        else if(e.getSource() == m_browseButton)
```

```
        {
            JFileChooser fc = new JFileChooser();
            fc.setCurrentDirectory(new File("../"));
            int returnVal = fc.showOpenDialog(this);
            if(returnVal == 0)
            {
                File file = fc.getSelectedFile();
                m_fileName.setText(file.getPath());
            }
        }
    }
}
}
```

5.11. Compiling ExampleDelayTest

The source code of this chapter can be compiled with the following.

Code 33. Compiling ExampleDelayTest

```
javac "ExampleDelayTest\src\recordingblogs\com\AudioBuffer.java" -d
    "ExampleDelayTest\class" -classpath
    "ExampleDelayTest\class;Orange\oreffect.jar"
javac "ExampleDelayTest\src\recordingblogs\com\WaveFile.java" -d
    "ExampleDelayTest\class" -classpath
    "ExampleDelayTest\class;Orange\oreffect.jar"
javac "ExampleDelayTest\src\recordingblogs\com\SelectDialog.java" -d
    "ExampleDelayTest\class" -classpath
    "ExampleDelayTest\class;Orange\oreffect.jar"
javac "ExampleDelayTest\src\recordingblogs\com\WaveFileDialog.java" -d
    "ExampleDelayTest\class" -classpath
    "ExampleDelayTest\class;Orange\oreffect.jar"
javac "ExampleDelayTest\src\recordingblogs\com\EffectStoreItem.java" -d
    "ExampleDelayTest\class" -classpath
    "ExampleDelayTest\class;Orange\oreffect.jar"
javac "ExampleDelayTest\src\recordingblogs\com\EffectStore.java" -d
    "ExampleDelayTest\class" -classpath
    "ExampleDelayTest\class;Orange\oreffect.jar"
javac "ExampleDelayTest\src\recordingblogs\com\Mixer.java" -d
    "ExampleDelayTest\class" -classpath
    "ExampleDelayTest\class;Orange\oreffect.jar"
javac "ExampleDelayTest\src\recordingblogs\com\EffectDialog.java" -d
    "ExampleDelayTest\class" -classpath
    "ExampleDelayTest\class;Orange\oreffect.jar"
javac "ExampleDelayTest\src\recordingblogs\com\MainFrame.java" -d
    "ExampleDelayTest\class" -classpath
    "ExampleDelayTest\class;Orange\oreffect.jar"
javac "ExampleDelayTest\src\recordingblogs\com\ExampleDelayTest.java" -d
    "ExampleDelayTest\class" -classpath
    "ExampleDelayTest\class;Orange\oreffect.jar"
```

```
jar cfm ExampleDelayTest\exampledelaytest.jar ExampleDelayTest\manifest.txt -C
    ExampleDelayTest\class .
```

oreffect.jar is needed for this compilation and for execution.

The manifest noted in the last line above should be used to specify the main class of the package and thereby make the JAR file executable. Its contents can be as follows.

Code 34. Manifest for ExampleDelayTest

```
Manifest-Version: 1.0
Class-Path: . oreffect.jar
Main-Class: recordingblogs.com.ExampleDelayTest
Created-By: Recordingblogs.com
Implementation-Title: "ExampleDelayTest"
Implementation-Version: "3.0.0"
Implementation-Vendor: "RecordingBlogs.com"
```

Chapter 6. Delay

A simple *delay* introduces a single repetition of the signal with some delay in time and some decay in amplitude. It is also called a *feedforward comb filter*. Feedforward comb filters are described in chapter 15 of volume 1.

6.1. Implementation of the delay

The following is an implementation of the delay in the Orinj framework, assuming one channel of audio with the sampling resolution of 16 bits. The graphical user interface for this delay is described in the sections that follow.

Code 35. Delay

```java
package mycompany.com;

import java.io.*;
import java.util.Vector;
import javax.sound.sampled.AudioFormat;
import recordingblogs.com.oreffect.*;

public class Delay extends Undo implements EffectInterface
{
    public static final float MINDELAY = 0;
    public static final float MAXDELAY = 1;
    public static final float MINDECAY = 0;
    public static final float MAXDECAY = 1;

    private float m_delay;
    private float m_decay;

    private transient Vector<byte []> m_storeBuffers;

    public Delay()
    {
        m_delay = 0.4F;
        m_decay = 0.6F;
        m_storeBuffers = new Vector<byte []>(0);
    }

    public float getDelay()
    {
        return m_delay;
    }

    public void setDelay(float delay)
    {
        m_delay = Math.min(MAXDELAY, Math.max(MINDELAY, delay));
```

```java
    }

    public float getDecay()
    {
        return m_decay;
    }

    public void setDecay(float decay)
    {
        m_decay = Math.min(MAXDECAY, Math.max(MINDECAY, decay));
    }

    public boolean allowsDryWetMix()
    {
        return true;
    }

    public boolean allowsSideChaining()
    {
        return false;
    }

    public void startPlay()
    {
        m_storeBuffers.setSize(0);
    }

    public void stopPlay()
    {
    }

    public boolean hasData()
    {
        if (m_storeBuffers.size() > 0)
            return true;
        return false;
    }

    public void setLanguage(String language)
    {
    }

    public boolean writeObject(WriteInterface ar)
    {
        try
        {
            ar.writeFloat(m_delay);
```

```java
         ar.writeFloat(m_decay);
      }
      catch (IOException e)
      {
         return false;
      }

      return true;
   }

   public boolean readObject(ReadInterface ar)
   {
      try
      {
         m_delay = ar.readFloat();
         m_decay = ar.readFloat();
      }
      catch (IOException e)
      {
         return false;
      }

      return true;
   }

   public void setEqual(EffectInterface effect)
   {
      if (this == effect)
         return;
      if (effect.getClass() != this.getClass())
         return;
      Delay e = (Delay) effect;
      m_delay = e.m_delay;
      m_decay = e.m_decay;
   }

   public void apply(byte [] dry, byte [] wet, byte [] control,
      AudioFormat format, double time)
   {
      // Initialize variables (as in the distortion effect)
      int blockAlign = (format.getChannels() * format.getSampleSizeInBits()) / 8;
      int sFrom = 0;
      int sTo = 0;

      // Store incoming audio data.  These data may be used during this call to
      // the function, but may also be used in subsequent calls.  This depends on
      // the size of audio buffers
      byte [] storeBuffer = new byte [dry.length];
```

```
System.arraycopy(dry, 0, storeBuffer, 0, dry.length);
m_storeBuffers.add(storeBuffer);

// The effect must produce a delayed repetition of the signal and place
// that repetition in the wet buffer.  The current value of the wet buffer
// is equal to some past value in the incoming data with some decay.
// Compute where the first value of wet buffer should be taken from (how
// far back the delay needs to look).  The code starts with the first
// available buffer in m_storeBuffers - as far back as possible - and
// computes the actual byte position from the amount of delay in m_delay

// We compute this information once per call to the apply function.  We
// could do so only once in startPlay, but then the user will not be able
// to change the delay during playback.  We could do this many times in
// the apply function (i.e., for each sample of audio data), but that is
// probably too many computations.  As the code is now, if the user makes
// changes to the delay, they will be heard in the playback whenever the
// next call to the apply function occurs.  How often the call to apply
// occurs depends on how large the audio buffers are.  The larger the
// buffers, the slower the playback response is to user changes

// Start as far back as possible
int curBuffer = m_storeBuffers.size() - 1;

// Convert the amount of delay from seconds to bytes
int curByte = (int) (m_delay * format.getSampleRate() * blockAlign);

// Make sure that the amount of delay in bytes is aligned to blocks,
// otherwise the data will be meaningless
while (curByte > 0 && curByte % blockAlign != 0)
    curByte--;
curByte = Math.max(curByte, 0);

// Convert to a negative number just because of the following piece of code
curByte = -curByte;

// Find the right buffer and byte
while (curByte < 0)
{
    curByte += dry.length;
    curBuffer--;
}

// Pick up the right buffer if available.  (It will not be available at the
// beginning of playback)
byte bufferFrom[] = null;
if(curBuffer >= 0)
```

```
      bufferFrom = m_storeBuffers.get(curBuffer));

  // Check how far back into the original data the delay should look and drop
  // the stored incoming audio buffers that are no longer necessary.  Note
  // that the user may change the amount of delay during playback and these
  // dropped buffers may become necessary again.  They are, however, gone.
  // In this case, the effect will simply produce a pause in the wet signal
  // (but not in the original signal) and continue.  An alternative would be
  // to drop only buffers that are beyond the maximum amount of allowed delay
  while (curBuffer > 0)
  {
      m_storeBuffers.remove(0);
      curBuffer--;
  }


  // Compute the values in the wet signal.  This loop only works if the
  // current buffer in the stored buffers (curBuffer) is nonnegative.
  // Otherwise, we are looking to far in the past and we do not have enough
  // stored data, which will happen at the beginning of playback and may
  // happen if the user increases the size of the delay during playback.  In
  // this case, the wet signal will simply contain silence.
  for(int i = 0; i < wet.length; curByte += blockAlign, i += blockAlign)
  {
      if (curBuffer >= 0)
      {
          // Adjust the current index of the stored buffer, if needed.  This
          // happens as we move from one stored buffer to the next
          if(curByte >= bufferFrom.length)
          {
              curBuffer++;
              bufferFrom = m_storeBuffers.get(curBuffer));
              curByte = 0;
          }

          // Compute the delay and decay and place the result in the wet buffer
          sFrom = (short) (((bufferFrom[curByte + 1] & 0xff) << 8)
              + (bufferFrom[curByte] & 0xff));
          sTo = (int) Math.min(Math.max((float) sFrom * m_decay,
              Short.MIN_VALUE), Short.MAX_VALUE);

          wet[i] = (byte)(sTo & 0xff);
          wet[i + 1] = (byte)(sTo >>> 8 & 0xff);
      }
      else
      {
          // Put silence in the wet buffer
          wet[i] = (byte) 0;
          wet[i + 1] = (byte) 0;
```

```
                    }
              }
        }
}
```

The following is a description of selected member data and functions in this class.

float MINDELAY – This is the minimum delay allowed in the delay effect. The Orinj Delay allows a delay in the repeated signal between 0 seconds and 1 second. Parametrizing these minimum and maximum numbers is a good idea. It helps with error checking, the design of the user interface, and so on.

float MAXDELAY – The maximum allowed delay is 1 second.

float MINDECAY – The minimum allowed decay is zero. We use decay between 0 (full decay that zeroes out the repeated signal) and 1 (no decay).

float MAXDECAY – The maximum allowed decay is 1.

float m_delay – This is the amount of delay specified by the user, in seconds.

float m_decay – This is the amount of decay specified by the user.

Vector<byte []> m_storeBuffers – This member data item stores past audio data. This is quite important. Audio data tend to be large and is therefore sent to DSP effects in pieces. Almost all DSP effects must use past audio data and the audio data pieces must be stored somewhere. The next subsection below explains this further.

Delay() – This is the default and only constructor. All Orinj effects have a single constructor that is noted in the effect package XML file so that it can be instantiated by Orinj.

float getDelay() – This function provides access to the amount of delay **m_delay** to the graphical user interface.

void setDelay(float delay) – This function similarly provides access to the amount of delay **m_delay** to the graphical user interface. It allows the graphical user interface to set **m_delay**. As implemented above, the function does not include any error checking. Error checking is discussed below.

float getDecay() – This function provides access to the amount of decay **m_decay** to the graphical user interface.

void setDecay(float decay) – This function similarly provides access to the amount of delay **m_decay** to the graphical user interface.

void startPlay() – In this effect, **startPlay** removes any data in **m_storeBuffers**. Data may have remained in **m_storeBuffers** from previous plays. If these data are not removed, they may be heard in the current playback.

boolean hasData() – Even when the original signal has ended (i.e., at the end of the track or at the end of a wave in the specific track), the delay may still produce the delayed repetition of the original signal. As shown below in the implementation of the function **apply**, this information is stored in **m_storeBuffers** and is removed when no longer needed. Thus, this function returns **true** if there is information in **m_storeBuffers**. It is always safe to return **false** here, although that would mean that the repeated delayed signal will stop with the end of the original signal and any residual delayed repetitions after the end of the original signal will be ignored.

void apply(byte [] dry, byte [] wet, byte [] control, AudioFormat format, double time) – As with all effects throughout this book, this implementation only works for the sampling resolution of 16 bits. The effect is also implemented for single channel (mono) audio. A more practical delay will allow for different delay and decay amounts in the left and right channel for a more interesting effect. This is also briefly discussed below. Still other delays may have multiple repetitions and delay and decay sweeps, such as the echo, chorus, and multitap delay discussed in the chapters that follow.

6.2. Using past audio buffers

All effects described in this book, except for the distortion of the previous chapter, need to look into and use past audio data and require **m_storeBuffers**. Some effects do this more than once. **m_storeBuffers** stores incoming audio buffers and discards them, when they are no longer needed. **m_storeBuffers** was not used in the distortion, because the distortion worked only with the current audio buffer.

The alternative to creating, storing, and throwing away audio buffers is to create a large enough buffer or structure that can hold the maximum amount of delay. We can, for example, set up several buffers and cycle through them. This will avoid Java's garbage collection. For the sake of simplicity, however, we will continue to use the construction above. There are several pieces to that construction that are repeated in the effects below. These are explained above.

The following code snipped stores audio data for later use.

Code 36. Storing audio data

```
byte [] storeBuffer = new byte [dry.length];
System.arraycopy(dry, 0, storeBuffer, 0, dry.length);
m_storeBuffers.add(storeBuffer);
```

The following code finds the right buffer and byte based on the amount of delay.

Code 37. Finding the right audio data

```
int curBuffer = m_storeBuffers.size() - 1;
int curByte = (int) (m_delay * format.getSampleRate() * blockAlign);
while (curByte > 0 && curByte % blockAlign != 0)
   curByte--;
curByte = Math.max(curByte, 0);
curByte = -curByte;
```

```
while (curByte < 0)
{
   curByte += dry.length;
   curBuffer--;
}
byte bufferFrom[] = null;
if(curBuffer >= 0)
   bufferFrom = m_storeBuffers.get(curBuffer));
```

The following code snipped removes unused audio data.

Code 38. Removing unused past audio data

```
while (curBuffer > 0)
{
   m_storeBuffers.remove(0);
   curBuffer--;
}
```

The following code snipped increments the past audio buffer as we process samples.

Code 39. Incrementing buffer indices of past audio data

```
if(curByte >= bufferFrom.length)
{
   curBuffer++;
   bufferFrom = m_storeBuffers.get(curBuffer));
   curByte = 0;
}
```

6.3. Signal polarity

Many delay effects may sound better if one of the channels has inverted polarity (inverted phase). To invert the phase of the signal in the code above, we use **-sTo** instead of **sTo**.

6.4. Working with stereo data

The delay effect is much more interesting, when the right and left channel use independent and different amounts of delay and decay. This means that instead of using **m_delay** and **m_decay**, we will use **m_leftDelay**, **m_leftDecay**, **m_rightDelay**, and **m_rightDecay**. In the **apply** function, **curByte** and **curBuffer** could become **curByteLeft**, **curByteRight**, **curBufferLeft**, and **curBufferRight**. Note that the bytes for the left channel samples are at the beginning of the block align and the bytes for the right channel samples are at the end of the block align. **curByteLeft** and **curByteRight** must be aligned accordingly. The left and right channels could be processed completely independently or in the same loop as follows.

Code 40. The delay apply function for stereo audio

```
public void apply(byte [] dry, byte [] wet, byte [] control, AudioFormat format,
   double time)
```

```
{
    // Initialize variables
    int blockAlign = (format.getChannels() * format.getSampleSizeInBits()) / 8;
    int channels = format.getChannels();
    int sFrom = 0;
    int sTo = 0;

    // Store incoming audio data for later use
    byte [] storeBuffer = new byte [dry.length];
    System.arraycopy(dry, 0, storeBuffer, 0, dry.length);
    m_storeBuffers.add(storeBuffer);

    // Figure out the left channel delay parameters
    int curBufferLeft = m_storeBuffers.size() - 1;
    int curByteLeft = (int) (m_leftDelay * format.getSampleRate() * blockAlign);
    while (curByteLeft > 0 && curByteLeft % blockAlign != 0)
        curByteLeft--;
    curByteLeft = Math.max(curByteLeft, 0);
    curByteLeft = -curByteLeft;
    while (curByteLeft < 0)
    {
        curByteLeft += dry.length;
        curBufferLeft--;
    }
    byte bufferFromLeft[] = null;
    if(curBufferLeft >= 0)
        bufferFromLeft = m_storeBuffers.get(curBufferLeft));

    // Figure out the right channel delay parameters
    int curBufferRight = m_storeBuffers.size() - 1;
    int curByteRight = (int) (m_rightDelay * format.getSampleRate() * blockAlign);
    while (curByteRight > 0 && curByteRight % blockAlign != blockAlign / 2)
        curByteRight--;
    curByteRight = Math.max(curByteRight, blockAlign / 2);
    curByteRight = -curByteRight;
    while (curByteRight < 0)
    {
        curByteRight += dry.length;
        curBufferRight--;
    }
    byte bufferFromRight[] = null;
    if(curBufferRight >= 0)
        bufferFromRight = m_storeBuffers.get(curBufferRight));

    // Drop stored audio data that are not used by either of the channels
    while (curBufferLeft > 0 && curBufferRight > 0)
    {
        m_storeBuffers.remove(0);
```

```java
      curBufferLeft--;
      curBufferRight--;
   }

   // Apply the effect
   for(int i = 0; i < wet.length; curByteLeft += blockAlign,
      curByteRight += blockAlign, i += blockAlign)
   {
      // Apply the left channel delay effect
      if (curBufferLeft >= 0)
      {
         if(curByteLeft >= bufferFromLeft.length)
         {
            curBufferLeft++;
            bufferFromLeft = m_storeBuffers.get(curBufferLeft));
            curByteLeft = 0;
         }
         sFrom = (short)(((bufferFromLeft[curByteLeft + 1] & 0xff) << 8)
            + (bufferFromLeft[curByteLeft] & 0xff));
         sTo = (int)Math.min(Math.max((float) sFrom * m_leftDecay,
            Short.MIN_VALUE), Short.MAX_VALUE);
         wet[i] = (byte)(sTo & 0xff);
         wet[i + 1] = (byte)(sTo >>> 8 & 0xff);
      }
      else
      {
         wet[i] = (byte) 0;
         wet[i + 1] = (byte) 0;
      }

      // Apply the right channel delay effect
      if (channels > 1)
      {
         if (curBufferRight >= 0)
         {
            if(curByteRight >= bufferFromRight.length)
            {
               curBufferRight++;
               bufferFromRight = m_storeBuffers.get(curBufferRight));
               curByteRight = blockAlign / 2;
            }

            sFrom = (short) (((bufferFromRight[curByteRight + 1] & 0xff) << 8)
               + (bufferFromRight[curByteRight] & 0xff));
            sTo = (int)Math.min(Math.max((float) sFrom * m_rightDecay,
               Short.MIN_VALUE), Short.MAX_VALUE);
            wet[i + 2] = (byte)(sTo & 0xff);
```

```
            wet[i + 3] = (byte)(sTo >>> 8 & 0xff);
        }
        else
        {
            wet[i + 2] = (byte) 0;
            wet[i + 3] = (byte) 0;
        }
    }
  }
}
```

Effects must work separately with the left and right channels only when the user has access to separate parameters for the two channels. An echo, for example, may use different delays for the left and right channel, but a compressor is unlikely to use different thresholds and compression ratios.

6.5. Complex settings for the simple delay

With stereo data, there are two input signals, two repetitions with two time delays and two amplitude decays, and two output signals. A simple delay does not have to always use the same amount of delay and decay in both the left and right channels. With significantly different amounts, we can create, for example, the perception of a signal repetition that bounces from one of the channels to the other channel. Suppose that we introduce a 0.4 second delay and 0.8 decay in the left channel and an 0.8 second delay and 0.6 decay in the right channel. The 400 ms difference between the delays in both channels is large enough to create two separate sound repetitions, since the human ear would typically interpret anything over 100 ms as separate. The longer delay is combined with a larger decay, creating the perception of a decaying echo. The decays and delays are approximately related, as the signal amplitude is reduced by 20% for every 400 ms delay (although the decay in a realistic echo should probably be exponential). We have thus created a two-repetition echo, one that jumps from the middle, if the original signal is in the middle, to the left and then to the right.

Suppose that we are in a good size room or a hall, but not necessarily in the middle. Sound will bounce from the walls and come back, perhaps at different times and perhaps with different amplitude, as in the bounce delay above. If the room is very large and we are far from the middle, the difference in the time delays of the reflected sound would be very noticeable. The difference in decay will probably not be as noticeable, unless the walls are markedly different. In a smaller room, both the difference in delay and the difference in decay would be unnoticeable. To simulate this situation, we can use a delay of 0.3 seconds and a decay of 0.4 in the left channel and a delay of 0.4 seconds and a decay of 0.4 in the right channel. The difference in time delay between the left and right channel is a barely noticeable. There is no difference between the two decays. The result will be a hint of some spatial positioning.

A *slapback delay* is one which uses a very small delay in time – less than 20 ms – and creates the perception of a phased signal. If the delay in time in the slapback delay is small – 5 ms – and there is little decay, there is a perception of sound presence. That is, the sound is louder and perceived as being closer. Again, since we usually have a stereo recording with two channels, we could make this effect even more interesting if we change the left and right presence perception. To do so, we can introduce a delay of 5 ms in the right channel with no decay, and we use no delay or decay in the left channel.

Finally, a delay can be used to simulate a *chorus*. We use delays of 30 ms and 12 ms in the left and right channels respectively, with no decay. Although a real chorus will have multiple repetitions and will allow at least for delay sweeps, this simple delay may just sound close enough to a chorus. We will revisit delay sweeps and the chorus in chapter 9.

> Subsequent sections of this book do not discuss stereo audio. In all effects that follow, stereo audio can be implemented similarly, by treating each channel separately.

6.6. Implementation of the delay graphical user interface

The following is an implementation of the graphical user interface for the (single channel) delay in the Orinj framework.

Code 41. Delay graphical user interface

```java
package mycompany.com;

import java.awt.*;
import java.awt.event.*;
import java.util.Hashtable;
import javax.sound.sampled.AudioFormat;
import javax.swing.*;
import javax.swing.event.*;
import recordingblogs.com.oreffect.*;

public class DelayPanel extends JPanel implements EffectPanelInterface,
    ActionListener, FocusListener, ChangeListener
{
    private JTextField m_delayField;
    private JTextField m_decayField;
    private JSlider m_delaySlider;
    private JSlider m_decaySlider;

    private Delay m_delay;

    public class SliderLabel extends JLabel
    {
        public SliderLabel(String string)
        {
            super(string);
```

```java
         setFont(EffectFont.SMALLFONT);
         EffectForeground.setForeground(this);
      }
   }

   private class FocusPolicy extends ContainerOrderFocusTraversalPolicy
   {
      public Component getDefaultComponent(Container c)
      {
         return m_delayField;
      }

      public Component getLastComponent(Container c)
      {
         return m_decaySlider;
      }

      public Component getFirstComponent(Container c)
      {
         return m_delayField;
      }

      public Component getComponentBefore(Container c, Component a)
      {
         if(a == m_decaySlider)
            return m_decayField;
         if(a == m_decayField)
            return m_delaySlider;
         if(a == m_delaySlider)
            return m_delayField;
         if (a == m_delayField)
            return m_decaySlider;
         return m_delayField;
      }

      public Component getComponentAfter(Container c, Component a)
      {
         if(a == m_delayField)
            return m_delaySlider;
         if(a == m_delaySlider)
            return m_decayField;
         if(a == m_decayField)
            return m_decaySlider;
         if(a == m_decaySlider)
            return m_delayField;
         return m_delayField;
      }
   }
```

```
public DelayPanel(Delay delay, AudioFormat format)
{
   GridBagLayout layout = new GridBagLayout();
   GridBagConstraints constraints = new GridBagConstraints();
   setLayout(layout);

   JLabel delayLabel = new JLabel("Delay (ms):");
   delayLabel.setFont(EffectFont.LARGEFONT);
   constraints.gridx = 0;
   constraints.gridy = 0;
   constraints.weightx = 1;
   constraints.weighty = 0;
   constraints.fill = GridBagConstraints.VERTICAL;
   constraints.anchor = GridBagConstraints.NORTHWEST;
   layout.setConstraints(delayLabel, constraints);
   add(delayLabel);

   m_delayField = new JTextField(4);
   m_delayField.setToolTipText("Set the amount of delay");
   m_delayField.addActionListener(this);
   m_delayField.addFocusListener(this);
   constraints.gridy = 1;
   layout.setConstraints(m_delayField, constraints);
   add(m_delayField);

   // Delay values are displayed in milliseconds, but stored in seconds
   Hashtable<Integer, SliderLabel> numbersDelay = new Hashtable<Integer,
       SliderLabel>();
   numbersDelay.put(new Integer(0), new SliderLabel("0"));
   numbersDelay.put(new Integer(200), new SliderLabel("200"));
   numbersDelay.put(new Integer(400), new SliderLabel("400"));
   numbersDelay.put(new Integer(600), new SliderLabel("600"));
   numbersDelay.put(new Integer(800), new SliderLabel("800"));
   numbersDelay.put(new Integer(1000), new SliderLabel("1000"));

   m_delaySlider = new JSlider(JSlider.VERTICAL, (int) (Delay.MINDELAY *
       1000), (int) (Delay.MAXDELAY * 1000), 0);
   m_delaySlider.setMajorTickSpacing(200);
   m_delaySlider.setMinorTickSpacing(50);
   m_delaySlider.setPaintTicks(true);
   m_delaySlider.setPaintLabels(true);
   m_delaySlider.setFont(EffectFont.SMALLFONT);
   m_delaySlider.setLabelTable(numbersDelay);
   m_delaySlider.setToolTipText("Set the amount of delay");
   m_delaySlider.addChangeListener(this);
   m_delaySlider.addFocusListener(this);
```

```java
constraints.gridy = 3;
constraints.weighty = 1;
layout.setConstraints(m_delaySlider, constraints);
add(m_delaySlider);

JLabel decayLabel = new JLabel("Decay (%):");
decayLabel.setFont(EffectFont.LARGEFONT);
constraints.gridx = 1;
constraints.gridy = 0;
constraints.insets.left = 5;
constraints.weighty = 0;
layout.setConstraints(decayLabel, constraints);
add(decayLabel);

m_decayField = new JTextField(4);
m_decayField.setToolTipText("Set the amount of decay");
m_decayField.addActionListener(this);
m_decayField.addFocusListener(this);
constraints.gridy = 1;
layout.setConstraints(m_decayField, constraints);
add(m_decayField);

// Decay value are displayed in percentages, but stored as fractions
Hashtable<Integer, SliderLabel> numbersDecay = new Hashtable<Integer,
    SliderLabel>();
numbersDecay.put(new Integer(0), new SliderLabel("0"));
numbersDecay.put(new Integer(20), new SliderLabel("20"));
numbersDecay.put(new Integer(40), new SliderLabel("40"));
numbersDecay.put(new Integer(60), new SliderLabel("60"));
numbersDecay.put(new Integer(80), new SliderLabel("80"));
numbersDecay.put(new Integer(100), new SliderLabel("100"));

m_decaySlider = new JSlider(JSlider.VERTICAL, (int) (Delay.MINDECAY * 100),
    (int) (Delay.MAXDECAY * 100), 100);
m_decaySlider.setMajorTickSpacing(20);
m_decaySlider.setMinorTickSpacing(5);
m_decaySlider.setPaintTicks(true);
m_decaySlider.setPaintLabels(true);
m_decaySlider.setFont(EffectFont.SMALLFONT);
m_decaySlider.setLabelTable(numbersDecay);
m_decaySlider.setToolTipText("Set the amount of decay");
m_decaySlider.addChangeListener(this);
m_decaySlider.addFocusListener(this);
constraints.gridy = 3;
constraints.weighty = 1;
layout.setConstraints(m_decaySlider, constraints);
add(m_decaySlider);
```

```
      m_delay = delay;

      setFocusTraversalPolicy(new FocusPolicy());

      updateData();
   }

   public void stateChanged(ChangeEvent e)
   {
      if(e.getSource() == m_delaySlider)
      {
         m_delay.setDelay((float) m_delaySlider.getValue() / 1000F, false);
         updateData();
      }
      else if(e.getSource() == m_decaySlider)
      {
         m_delay.setDecay((float) m_decaySlider.getValue() / 100F, false);
         updateData();
      }
   }

   public void actionPerformed(ActionEvent e)
   {
      if(e.getSource() == m_delaySlider)
      {
         m_delay.setDelay((float) m_delaySlider.getValue() / 1000F, false);
         updateData();
      }
      else if(e.getSource() == m_decaySlider)
      {
         m_delay.setDecay((float) m_decaySlider.getValue() / 100F, false);
         updateData();
      }
   }

   public void focusGained(FocusEvent e)
   {
   }

   public void focusLost(FocusEvent e)
   {
      if(e.getSource() == m_delaySlider)
      {
         m_delay.setDelay((float) m_delaySlider.getValue() / 1000F, false);
         updateData();
      }
      else if(e.getSource() == m_decaySlider)
```

```
        {
            m_delay.setDecay((float) m_decaySlider.getValue() / 100F, false);
            updateData();
        }
        else if(e.getSource() == m_delaySlider)
        {
            m_delay.setDelay((float) m_delaySlider.getValue() / 1000F, true);
        }
        else if(e.getSource() == m_decaySlider)
        {
            m_delay.setDecay((float) m_decaySlider.getValue() / 100F, true);
        }
    }

    public void updateData()
    {
        m_delaySlider.setValue((int) (m_delay.getDelay() * 1000F));
        m_delayField.setText(Integer.toString((int) (m_delay.getDelay() * 1000F)));
        m_decaySlider.setValue((int) (m_delay.getDecay() * 100F));
        m_decayField.setText(Integer.toString((int) (m_delay.getDecay() * 100F)));
    }
}
```

The following describes selected member data and functions of this class.

JTextField m_delayField – This text field allows the user to specify the amount of delay. Note the choice to display the delay in milliseconds and the fact that **m_delay** in class **Delay** is in seconds. When converting from one to the other, the code multiplies or divides by 1000.

JTextField m_decayField – This text field allows the user to specify the amount of decay. As used here, for example, 80% means that 80% of the amplitude of the original signal is preserved in the delay repetition and 20% is decayed. Note that the delay is displayed here in percentages, whereas **m_decay** in class **Delay** is between 0 and 1. When converting from one to the other, the code multiplies or divides by 100.

JSlider m_delaySlider – This slider allows the user to specify the amount of delay. As with **m_delayField**, this slider displays information in milliseconds rather than seconds. Sliders are easier to use than text fields, as they require fewer mouse and key clicks by the user. Text fields, however, are more precise. Therefore, both are used in this effect. This means that the graphical user interface must adjust one as the other one changes. This is done primarily through the function **updateData**.

JSlider m_decaySlider – This slider allows the user to specify the amount of decay. Similarly to the slider above, this one works together with the text field **m_decayField** to set the decay.

Delay m_delay – This is the delay effect.

The code for the graphical user interfaces of the remaining effects in this book is not included. A user interface is needed for each effect, but is relatively easy to implement.

6.7. Creating presets in Orinj

A preset is a set of predetermined values for the user controls in an effect. Note that the delay and decay of the simple delay above are initialized to 0.4 and 0.6 respectively. The user can adjust these, of course, but it may be a good idea to suggest some delay and decay values to the user.

For example, the preset "bounce (quick)" for the Orinj Delay specifies that the delay in the left channel is 400 ms, the delay in the right channel is 800 ms, the decay in the left channel is 80%, and the decay in the right channel is 60%. Because of the different values for the delay and the decay in the left and right channel, the sound seems to bounce quickly between the left and the right.

Presets are simply a way of allowing the users to quickly find good settings for each effect, without having to try many options by themselves. The user is free to deviate from these, but at least there are some interesting effect values to start with.

Presets can be created in Orinj. Once a working effect is added to an Orinj track, the dialog that contains the effect controls also lists available presets and allows the user to save additional presets.

Presets are saved in binary files using the **writeObject** function of **EffectInterface** and are read from these files using the **readObject** function of **EffectInterface**. The names of these files carry the name of the effect and the name of the effect class. For example, the Orinj preset "bounce (quick).recordingblogs.com.oreffectpack.delay" is the preset "bounce (quick)" for the Orinj Delay effect of the package **recordingblogs.com.oreffectpack**. Presets are saved in the "orange/presets" folder of the Orinj installation. Any effect package installations should place the effect package in the "orange/effect" folder of the Orinj installation and the effect presets in the "orange/presets" folder of the installation.

Presets are not required. The Orinj phase oscilloscope and spectrogram for example simply monitor audio data and do not have user controls. Therefore, they do not have presets.

For some effects, such as the distortion in chapter 4, presets may not be useful. The distortion sound depends heavily on the level of the incoming signal, relative to the distortion threshold. Since the levels of incoming signals vary, the user must adjust the threshold accordingly, independently of what threshold level is suggested by a preset.

Chapter 7. Echo

An *echo* introduces multiple repetitions of the signal, usually with the same amount of delay and decay between each successive repetition. An echo is a *feedback comb filter*. Feedback comb filters are described in chapter 15 of volume 1.

The only difference between the feedback comb filter (the echo) and the feedforward comb filter (the simple delay of the previous chapter) is that the feedforward comb filter takes a repetition of the original signal, whereas the feedback comb filter takes a repetition of the repeated signal. By feeding the repetition through the comb filter again and again, the effect creates successive repetitions with an ever-increasing delay and an ever-decreasing amplitude.

It would be easy to create an echo by simply taking the code for the delay, but instead of storing and using the original (dry) signal in **m_storeBuffers**, storing and using the repeated (wet) signal. This is one possible and more efficient implementation.

We use a different approach. We compute each successive repetition. This is less efficient, but allows control over the total number of repetitions. An echo will decay to silence sooner or later. However, if there are too many repetitions to the echo, the echo may just be too much. It may be good to allow the user to limit the total number of repetitions.

The member data of the echo are the same as that for the delay – **MINDELAY**, **MAXDELAY**, **MINDECAY**, **MAXDECAY**, **m_delay**, **m_decay**, and **m_storeBuffers** – with the addition of the following items.

int MIN_REPETITIONS – This is the minimum allowed value for **m_maxRepetitions**, the maximum number of echo repetitions. An echo on a mono signal with no repetitions is not useful, but in stereo data a user may opt for introducing no repetitions in one channel and an echo in the other.

int MAX_REPETITIONS – This is the maximum allowed value for **m_maxRepetitions**, the maximum number of echo repetitions.

int m_maxRepetitions – The maximum number of echo repetitions allowed by the user.

int [] m_decays – The value of the decay used for each repetition. The length of this array can be set to **MAX_REPETITIONS** in the effect constructor or on **startPlay**.

int [] m_curBuffer – This variable keeps track of which of the stored past audio buffers are used by the echo repetitions. In the delay of the previous chapter, this variable was declared inside the **apply** function. Here, this variable is an array that handles multiple repetitions. It is declared as member data of the class so that the array does not have to be recreated on each call to the **apply** function.

int [] m_curByte – This variable keeps track of which bytes in the stored past audio buffers are used by the echo repetitions. Similarly to **m_curBuffer**, this variable is member data of the class so that the array is not recreated on each call to the **apply** function.

The constructor of the echo is as follows.

Code 42. Echo constructor

```
public Echo()
{
   // A delay of zero and decay of 1 is not useful.  Maximum repetitions of 1 is
   // not an echo, but is a simple delay.  The user must adjust these parameters
   // to get an interesting echo
   m_delay = 0.0F;
   m_decay = 1.0F;
   m_maxRepetitions = 1;

   // Initialize other parameters
   m_curBuffer = new int[MAX_REPETITIONS];
   m_curByte = new int[MAX_REPETITIONS];
   m_decays = new float[MAX_REPETITIONS];
   m_storeBuffers = new Vector<byte []>(0);
}
```

The following code is executed at the beginning of playback.

Code 43. Echo at the beginning of playback

```
public void startPlay()
{
   // This implementation is the same as the one for the delay
   m_storeBuffers.setSize(0);
}
```

The following is the implementation of the echo.

Code 44. Echo

```
public void apply(byte [] dry, byte [] wet, byte [] control, AudioFormat format,
   double time)
{
   // Compute the decay for each successive repetition
   for(int i = 0; i < m_maxRepetitions; i++)
      m_decays[i] = (float) Math.pow(m_decay, i + 1);

   // Initialize variables
   int blockAlign = (format.getChannels() * format.getSampleSizeInBits()) / 8;
   int channels = format.getChannels();
   int sFrom = 0;
   int sTo = 0;
   byte bufferFrom[] = null;

   // The following is technically an unnecessary variable that can be computed
   // from the longest echo repetition.  This variable will become useful if
```

```
// there are more than one channels and each channel uses a different delay
// amount of delay.  This variable keeps track of which stored past audio
// buffers are used
int firstUsedBuffer = 0;

// Store the incoming audio data for later use
byte [] storeBuffer = new byte [dry.length];
System.arraycopy(dry, 0, storeBuffer, 0, dry.length);
m_storeBuffers.add(storeBuffer);

// As in the delay, compute how far back the echo should look for each
// repetition
for(int i = 0; i < m_maxRepetitions; i++)
{
    m_curBuffer[i] = m_storeBuffers.size() - 1;
    m_curByte[i] = (int) (m_delay * (i + 1) * format.getSampleRate() *
        blockAlign);
    while (m_curByte[i] > 0 && m_curByte[i] % blockAlign != 0)
        m_curByte[i]--;
    m_curByte[i] = Math.max(m_curByte[i], 0);
    m_curByte[i] = -m_curByte[i];
    while (m_curByte[i] < 0)
    {
        m_curByte[i] += dry.length;
        m_curBuffer[i]--;
    }

    // Keep track of which stored buffers are used, so that the rest can be
    // dropped
    if(i == 0)
        firstUsedBuffer = m_curBuffer[i];
    else
        firstUsedBuffer = Math.min(firstUsedBuffer, m_curBuffer[i]);
}

// Compute the echo by computing all repetitions with the delays and decays
// computed above
for(int i = 0; i < dry.length; i += blockAlign)
{
    sTo = 0;

    for (int j = 0; j < m_maxRepetitions; j++)
    {
        if (m_curBuffer[j] >= 0)
        {
            bufferFrom = m_storeBuffers.get(m_curBuffer[j]);
            if (m_curByte[j] >= bufferFrom.length)
            {
```

```
                    m_curBuffer[j]++;
                    bufferFrom = m_storeBuffers.get(m_curBuffer[j]);
                    m_curByte[j] = 0;
                }
                sFrom = (short) (((bufferFrom[m_curByte[j] + 1] & 0xff) << 8)
                    + (bufferFrom[m_curByte[j]] & 0xff));
                sTo = (int) ((float) sTo + (float) sFrom * m_decays[j]);
            }

            m_curByte[j] += blockAlign;
        }

        sTo = Math.min(Math.max(sTo, Short.MIN_VALUE), Short.MAX_VALUE);
        wet[i] = (byte) (sTo & 0xff);
        wet[i + 1] = (byte) (sTo >>> 8 & 0xff);
    }

    // Remove unnecessary stored buffers
    while (firstUsedBuffer > 0)
    {
        m_storeBuffers.remove(0);
        firstUsedBuffer--;
    }
}
```

As with the delay, the echo is more interesting if the user can employ different amounts of delay and decay in each channel. Inverting the phase of the signal is also possible.

An implementation of the echo graphical user interface is not included in this book. The echo controls are the same as those of the delay, except for the addition of a control for the maximum number of echo repetitions.

Chapter 8. Multitap delay

A *multitap delay* is essentially a collection of multiple signal repetitions, where there is no relationship between the delays or decays of the repetitions. In its purest form, a multitap delay is a *tapped delay line* like the one discussed in chapter 15 of volume 1. Depending on the construction (e.g., the use of feedback), there may be some relationship between some signal repetitions unlike a simple tapped delay line.

Since we implemented an echo not by using a feedback comb filter, but by computing each separate repetition independently, we already have the algorithm to apply a multitap delay. We can use the algorithm of the echo if we know the delay and decay of each repetition.

The following diagram depicts the Orinj multitap delay.

Figure 4. Schema for a multitap delay

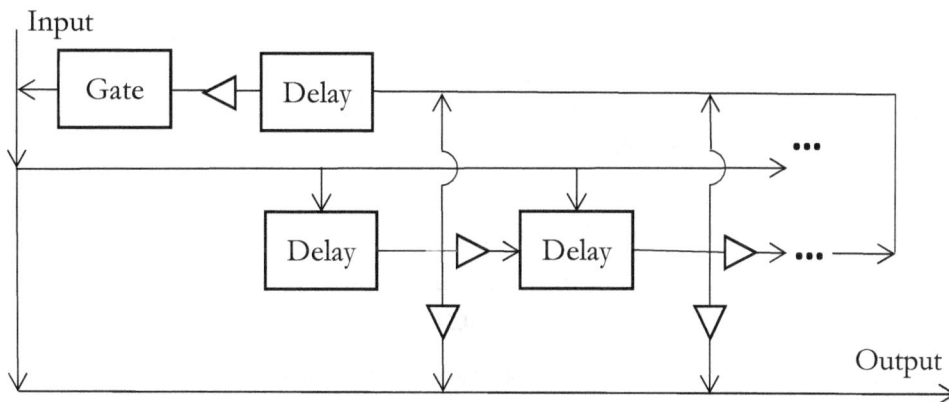

The input signal, besides going directly towards the output, is fed into a series of parallel, independent simple delays, each creating a single repetition with some delay in time and some decay in amplitude. The outputs of the simple delays are fed into output, perhaps with additional decay, and back towards the input, where they may be subject to an additional overall feedback simple delay and a gate.

Each small triangle represents gain applied to the signal – some decay. The second-row delays are the individual taps. The top-row delay is the delay applied to the feedback signal. The gate zeroes out the signal if the signal drops below a certain level, effectively limiting feedback. There are several ways to implement this gate. We can use an actual gate effect, but the simplest way is to limit the number of signal repetitions.

This multitap delay is made more complex by the fact that it allows feedback. While each of the taps is a simple delay on its own, the feedback can create multiple repetitions of taps and allows repetitions of a tap because of another tap. The code snippet below shows how to convert the diagram above into a set of delays and decays that can be used in the code for the echo effect.

To implement the diagram above, the Orinj multitap delay defines the following member data.

private Vector<Tap> m_taps – This is a collection of delays (taps). In the schematic above, these are represented by the "delay" rectangles in the second row of rectangles and the "decay" triangles that follow the delays. Each delay introduces a single repetition of the signal with some delay and some decay. This repetition is then fed into the next delay and back into the input.

float m_feedbackDelay – This is the amount of delay applied to the feedback signal. It is represented by the topmost delay rectangle.

float m_feedbackDecay – This is the amount of decay applied to the delayed feedback signal, represented by the topmost triangle in the schematic above.

int m_maxRepetitions – This is the maximum number of feedback repetitions, represented by the gate in the schematic above.

int MAXIMPULSES – This is the maximum number of total signal repetitions (also used as the effective signal gate).

float [] m_impulseDelay – These are the delays for all signal repetitions. They are computed with the code below from the parameters in the schematic above. These data and **m_impulseDelay** below are what we can use in the code for the echo to implement the multitap delay. This array is of size **MAXIMPULSES**.

float [] m_impulseDecay – These are the decays for all signal repetitions, computed with the code below from the parameters of the schematic above. This array is of size **MAXIMPULSES**.

int m_impulseSize – This is a counter of the computed number of impulse delays and decays. It is computed below and is between 0 and **MAXIMPULSES**.

A single delay (a tap), consistent with the schematic above, contains three parameters. Those are the three variables of the structure **Tap** below.

float m_delay – This is the delay introduced by the tap.

float m_tapDecay – This is the decay in the delayed repetition of the tap.

float m_feedDecay – This is the additional decay in the delayed repetition, introduced before this repetition is fed into the next tap.

The class Tap is as follows.

Code 45. A multitap delay tap

```
public class Tap
{
    private float m_delay;
    private float m_tapDecay;
    private float m_feedDecay;

    public Tap()
    {
```

```
        m_delay = 0.5F;
        m_tapDecay = 0.5F;
        m_feedDecay = 0.5F;
    }

    // A number of get() and set() functions should be implemented here
    ...
}
```

The following function computes the delays and decays (the impulses) of the multitap delay.

Code 46. Computing the impulses of a multitap delay

```
public void computeDelaysAndDecays()
{
    m_impulseSize = 0;

    if (getSize() < 1)
        return;

    float delay = 0;
    float decay = 0;
    int size = 0;

    // We omit repetitions with decays of less than -60 dB, which is
    // equivalent to amplitude of 0.001
    double MINIMUMDECAY = 0.001;

    // Compute the direct delay and decay of the first tap
    if (m_taps.get(0).getTapDecay() >= MINIMUMDECAY)
    {
        m_impulseDelay[m_impulseSize] = m_taps.get(0).getDelay();
        m_impulseDecay[m_impulseSize] = m_taps.get(0).getTapDecay();
        m_impulseSize++;
    }

    // Go through the remaining taps
    for(int i = 1; i < getSize(); i++)
    {
        // The input to each tap is the original signal plus the output of the
        // previous tap
        delay = m_taps.get(i - 1).getDelay();
        decay = m_taps.get(i - 1).getTapDecay();
        delay += m_taps.get(i).getDelay();
        decay *= m_taps.get(i).getTapDecay();
        decay *= m_taps.get(i - 1).getFeedDecay();
        if (decay >= MINIMUMDECAY && m_impulseSize < MAXIMPULSES)
        {
            m_impulseDelay[m_impulseSize] = delay;
```

```
            m_impulseDecay[m_impulseSize] = decay;
            m_impulseSize++;
        }
        if (m_taps.get(i).getTapDecay() >= MINIMUMDECAY
            && m_impulseSize < MAXIMPULSES)
        {
            m_impulseDelay[m_impulseSize] = m_taps.get(i).getDelay();
            m_impulseDecay[m_impulseSize] = m_taps.get(i).getTapDecay();
            m_impulseSize++;
        }
    }

    // We have created a pattern.  Feed this pattern back into the input
    // (the feedback) and delay and decay it
    int startFeedback = 0, endFeedback = m_impulseSize, startPreviousTap = 0;
    for(int i = 0; i < m_maxRepetitions; i++)
    {
        // Run the feedback through the first tap
        for(int j = startFeedback; j < endFeedback; j++)
        {
            delay = m_impulseDelay[j];
            decay = m_impulseDecay[j];
            delay += m_feedbackDelay;
            decay *= m_feedbackDecay;
            delay += m_taps.get(0).getDelay();
            decay *= m_taps.get(0).getTapDecay();
            if (decay >= MINIMUMDECAY && m_impulseSize < MAXIMPULSES)
            {
                m_impulseDelay[m_impulseSize] = delay;
                m_impulseDecay[m_impulseSize] = decay;
                m_impulseSize++;
            }
        }

        // Continue with the remaining taps
        for(int j = 1; j < getSize(); j++)
        {
            // Record this value for later
            startPreviousTap = m_impulseSize;

            // Run the whole feedback through the tap
            for(int k = startFeedback; k < endFeedback; k++)
            {
                delay = m_impulseDelay[k];
                decay = m_impulseDecay[k];
                delay += m_feedbackDelay;
                decay *= m_feedbackDecay;
```

```
                delay += m_taps.get(j).getDelay();
                decay *= m_taps.get(j).getTapDecay();
                if (decay >= MINIMUMDECAY && m_impulseSize < MAXIMPULSES)
                {
                    m_impulseDelay[m_impulseSize] = delay;
                    m_impulseDecay[m_impulseSize] = decay;
                    m_impulseSize++;
                }
            }

            // Run the previous tap through the taps
            size = m_impulseSize;
            for(int k = startPreviousTap; k < size; k++)
            {
                delay = m_impulseDelay[k];
                decay = m_impulseDecay[k];
                delay += m_feedbackDelay;
                decay *= m_feedbackDecay;
                delay += m_taps.get(j).getDelay();
                decay *= m_taps.get(j).getTapDecay();
                decay *= m_taps.get(j - 1).getFeedDecay();
                if (decay >= MINIMUMDECAY && m_impulseSize < MAXIMPULSES)
                {
                    m_impulseDelay[m_impulseSize] = delay;
                    m_impulseDecay[m_impulseSize] = decay;
                    m_impulseSize++;
                }
            }
        }

        startFeedback = endFeedback;
        endFeedback = m_impulseSize;
    }

    // Cleanup
    for(int i = m_impulseSize; i < MAXIMPULSES; i++)
    {
        m_impulseDelay[i] = 0;
        m_impulseDecay[i] = 0;
    }
}
```

The constructor of the effect is as follows.

Code 47. Multitap delay constructor

```
public MultitapDelay()
{
    // Create at least one tap
```

```
    m_taps = new Vector<Tap>(0);
    m_taps.add(new Tap());
    m_currentTap = 0;

    // Initialize global variables
    m_feedbackDelay = 0.5F;
    m_feedbackDecay = 0.5F;
    m_maxRepetitions = 5;
    m_maxImpulses = MAXIMPULSES;

    // Initialize storage of past audio data
    m_storeBuffers = new Vector<byte []>(0);

    // Initialize the delays and decays that are computed at the beginning of
    // playback
    m_impulseDelay = new float [MAXIMPULSES];
    m_impulseDecay = new float [MAXIMPULSES];
    m_impulseSize = 0;
    for (int i = 0; i < MAXIMPULSES; i++)
    {
        m_impulseDelay[i] = 0;
        m_impulseDecay[i] = 0;
    }

    // Initialize the arrays that keep track of how far back in past audio data
    // each repetition should look
    m_curBuffer = new int [MAXIMPULSES];
    m_curByte = new int [MAXIMPULSES];
}
```

The following code is executed at the beginning of playback.

Code 48. Multitap delay at the beginning of playback

```
public void startPlay()
{
    // Initialize the storage of past audio data
    m_storeBuffers.setSize(0);

    // Compute delays and decays for all repetitions
    computeImpulse();
}
```

The implementation of the multitap delay is as follows.

Code 49. Multitap delay

```
public void apply(byte [] dry, byte [] wet, byte [] control, AudioFormat format,
    double time)
```

```
{
   // Initialize variables
   int blockAlign = (format.getChannels() * format.getSampleSizeInBits()) / 8;
   int sFrom = 0;
   int sTo = 0;
   byte bufferFrom[] = null;
   int firstUsedBuffer = 0;

   // Store audio data for later use
   byte [] storeBuffer = new byte [dry.length];
   System.arraycopy(dry, 0, storeBuffer, 0, dry.length);
   m_storeBuffers.add(storeBuffer);

   // Compute how far back into past audio data each delay repetition should look
   for(int i = 0; i < m_impulseSize; i++)
   {
      m_curBuffer[i] = m_storeBuffers.size() - 1;
      m_curByte[i] = (int) (m_impulseDelay[i] * format.getSampleRate()
         * blockAlign);
      while (m_curByte[i] > 0 && m_curByte[i] % blockAlign != 0)
         m_curByte[i]--;
      m_curByte[i] = Math.max(m_curByte[i], 0);
      m_curByte[i] = -m_curByte[i];
      while (m_curByte[i] < 0)
      {
         m_curByte[i] += dry.length;
         m_curBuffer[i]--;
      }

      // Keep track of which buffers we are using so that we can remove those
      // not used
      if(i == 0)
         firstUsedBuffer = m_curBuffer[i];
      else
         firstUsedBuffer = Math.min(firstUsedBuffer, m_curBuffer[i]);
   }

   // Calculate the effect
   for(int i = 0; i < dry.length; i += blockAlign)
   {
      sTo = 0;

      // Add each of the repetitions
      for (int j = 0; j < m_impulseSize; j++)
      {
         if (m_curBuffer[j] >= 0)
         {
            bufferFrom = m_storeBuffers.get(m_curBuffer[j]);
```

```
                if (m_curByte[j] >= bufferFrom.length)
                {
                    m_curBuffer[j]++;
                    bufferFrom = m_storeBuffers.get(m_curBuffer[j]);
                    m_curByte[j] = 0;
                }

                sFrom = (short) (((bufferFrom[m_curByte[j] + 1] & 0xff) << 8)
                    + (bufferFrom[m_curByte[j]] & 0xff));
                sTo = (int) ((float) sTo + (float) sFrom * m_impulseDecay[j]);
            }

            m_curByte[j] += blockAlign;
        }

        // Store output
        sTo = Math.min(Math.max(sTo, Short.MIN_VALUE), Short.MAX_VALUE);
        wet[i] = (byte) (sTo & 0xff);
        wet[i + 1] = (byte) (sTo >>> 8 & 0xff);
    }

    // Remove unused buffers
    while (firstUsedBuffer > 0)
    {
        m_storeBuffers.remove(0);
        firstUsedBuffer--;
    }
}
```

Chapter 9. Chorus

A *chorus*, similarly to an echo or a multitap delay, adds several repetitions to the signal. The chorus repetitions, however, have the same amplitude as the original signal. There is no decay. The delays between the original signal and the repetitions tend to be very short – perhaps around 20 milliseconds. More importantly, to simulate the natural variations that may occur in a choir, the delays vary in time, becoming larger, then smaller, and then larger again. If the delays vary quickly, such as one to ten times per second from shortest to longest and back, the change in the underlying wave form of the repetition signal may result in a slight change in signal pitch.

Delays and echoes allow only constant amounts of delay and decay. In some practical cases, it is useful to introduce gradual changes in the amount of delay time (*delay sweeps*) or gradual changes in the amount of decay (*decay sweeps*). To simulate an actual choir, for example, we must account for the fact that choir members will never have the same timing or the same amplitude. In a good choir, the difference between one voice and another would be very small and the rate of change of that difference probably negligible, but it does exist. In a not so good choir, that difference may be very noticeable. Practical chorus effects would usually introduce only delay sweeps and not decay sweeps – we expect that in most uses of chorus effects amplitude differences and variations would be either unnecessary or unnoticeable.

A chorus can be implemented similarly to an echo or a multitap delay. The only difference is the variation in the repetition delays in time. We must be careful to allow for a gradual variation. If the variation is not gradual, it may result in discontinuities in the repetition signal that may produce audible pops in the sound.

We must also ensure that the delays, if there are more than one, and their variations are not the same so that the listener perceives the presence of multiple voices in the signal and the thickness of the sound that the chorus should produce, even if the listener is not able to clearly distinguish one voice from another.

The following are the member data of the chorus.

final static int MAXVOX – This is the maximum number of chorus voices (repetitions of the original signal) that the user can choose to have. The chorus allows up to 100 voices, but often we use choruses with much fewer voices, such as up to 10.

int m_vox – This is the number of chorus voices. The maximum is **MAXVOX**. There should be at least one repetition for an audible chorus, but we could allow zero repetitions if we work with more than one channels, so that one of the channels has no chorus (zero repetitions), while the other channel has a chorus (more repetitions).

float m_rate – This is the rate of variation in the delay of a repetition, measured in milliseconds per second. In other words, the delay between the original signal and the repetition changes, becoming larger or smaller with **m_rate** milliseconds for each second of playback. This is not the most standard way to define the chorus delay changes, but it works. Typically, the chorus modulation is defined as the number of periodic fluctuations in the chorus delay, from smaller to larger and back to smaller, measured in Hz. With the two parameters below, however, these two measures of the delay variations can be computed from each other.

float m_min – This is the minimum delay between a repetition and the original signal, measured in seconds.

float m_max – This is the maximum delay between a repetition and the original signal, measured in seconds.

double [] m_delayStart – This is the initial delay, at playback start, between the original signal and each of the repetitions (each of the voices), in seconds. Since these amounts are randomized below, it may be useful to record them to ensure that each playback is the same as the previous one. In Orinj, these amounts are stored as part of the effect serialization with **EffectInterface::writeObject**. To simplify things, an array of length **MAXVOX** can be allocated.

boolean m_delaySet – This is a flag to check during playback whether individual repetition delays have been set (**true**) or should be computed (**false**). This flag is necessary, as the user may change the number of voices, the rates of delay variation, or the minimum and maximum delays during playback.

double [] m_delay – This is the current value of delay for each of the voices, measured in seconds. These values start with **m_delayStart**, but vary according to **m_rates**. This array should also be of length **MAXVOX**.

double [] m_rates – These are the rates of variation in the delays between the original signal and each of the repetitions, in milliseconds per second. These are set closely to **m_rate**, but randomized nonetheless, and hence also stored similarly to **m_delayStart**. This array should also be of length **MAXVOX**.

boolean m_ratesSet – This is a flag to check during playback whether the rates of delay variations have been set for individual repetitions (**true**) or should be computed (**false**).

Vector<byte []> m_storeBuffers – as with all other effects in this book, these are stored buffers of past audio data.

int [] m_curBuffer – These are the indices of the buffers in **m_storeBuffers** that should be used by the effect to retrieve the audio data for the delayed repetition. As in the delay and echo in previous chapters, the chorus could compute this information on every call of its **apply** function, but we would like to avoid any possible gaps in the chorus repetitions and audible pops in the sound. Instead of computing this information at the beginning of every call to the **apply** function as was done in the delay and the echo, we do so once and recompute this information only if the user changes the parameters of the chorus. A similar approach can be used in the delay and the echo and may be beneficial, as it reduces the computations needed in the **apply** function. This array should also be of length **MAXVOX**.

int [] m_curByte – These are the indices of the bytes in the **m_storeBuffers** buffers that should be used by the effect to retrieve the audio data for the delayed repetition. This array should also be of length **MAXVOX**.

double [] m_accumulateCurByte – This variable accumulates information on how **m_curByte** should change as the repetition delay varies. Suppose that the delay between the original signal and a repetition in 44100 Hz sampled audio starts at 10 ms = 0.01 seconds = 441 samples and changes by 5 milliseconds per second (i.e., at the end of one second, the delay should become 15 ms = 0.015 seconds = 661 samples). In one second, we increase in the delay by 0.005 * 44100 = 220 samples. We cannot do that at once. We must do it gradually. For every audio sample, we add 220 / 44100 = 0.005 samples to the delay. In 1 / 0.005 = 200 samples, we add 1 sample to the delay. In computing the repetition, we repeat a byte from **m_storeBuffers** every 200 samples (if we were decreasing the delay, we would skip a byte). Repeating one byte does not cause major distortion in the signal and should not result in an audible pop in the sound. Jumping by several bytes (e.g., 200 bytes once per second), will cause an audible pop. This array should also be of length **MAXVOX**.

boolean m_curByteSet – This is a flag to check whether the effect knows how far back in past historical data it should look when computing the delayed signal (**true**) or whether it needs to compute this information (**false**).

The constructor of the chorus is as follows.

Code 50. Chorus constructor

```
public Chorus()
{
   // These four parameters are set by the user
   m_rate = 0.5F;
   m_vox = 20;
   m_min = 15;
   m_max = 100;

   // The initial delays and initial rates of delay variations are random,
   // between the values of m_min and m_max set by the user and close to m_rate,
   // but different from each other.  These are set at the beginning of playback
   m_delayStart = new double[MAXVOX];
   m_rates = new double[MAXVOX];
   for(int i = 0; i < MAXVOX; i++)
   {
     m_delayStart[i] = 0;
     m_rates[i] = 0;
   }

   // These delays start with the values of m_delayStart, but vary over time with
   // m_rates
   m_delay = new double[MAXVOX];
```

```
// Let the chorus know that delays and rates should be computed at the
// beginning of playback
m_delaySet = false;
m_ratesSet = false;
m_curByteSet = false;

// Keep track of past audio data for the delays
m_curByte = new int[MAXVOX];
m_curBuffer = new int[MAXVOX];
m_storeBuffers = new Vector<byte []>(0);

// Make sure that skipping bytes or repeating bytes happens slowly as the
// delays vary, so that there are no audible pops in the signal
m_accumulateCurByte = new double[MAXVOX];
}
```

The following code is executed at the beginning of playback.

Code 51. Chorus at the beginning of playback

```
public void startPlay()
{
    // Calculate initial delays
    if(!m_delaySet)
        computeDelayStart();

    // Set the rates of variation in delays
    if(!m_ratesSet)
        computeRates();

    // Make sure that the chorus computes how far back it should look
    // in the past audio data when applying the chorus delays
    m_curByteSet = false;

    // Initialize past audio storage
    m_storeBuffers.setSize(0);
}

// Compute the initial delays between the original signal and the chorus voices
// (the signal repetitions).  These delays are randomized inside the interval
// between m_min and m_max
public void computeDelayStart()
{
    for(int i = 0; i < MAXVOX; i++)
    {
        m_delayStart[i] = (m_min + ((m_max - m_min) / m_vox) * i) +
            (Math.random() * (m_max - m_min) / m_vox);
```

```
            m_delayStart[i] = Math.min(Math.max(m_delayStart[i], m_min), m_max);
            m_delay[i] = m_delayStart[i];
        }

    m_delaySet = true;
}


// Compute the rates of variation in the delays between the original signal and
// the signal repetitions.  Rates are randomized between zero and two times the
// value specified by the user in m_rate
public void computeRates()
{
    for(int i = 0; i < MAXVOX; i++)
        m_rates[i] = 2D * m_rate * Math.random();
    m_ratesSet = true;
}
```

The implementation of the chorus is as follows.

Code 52. Chorus

```
public void apply(byte [] dry, byte [] wet, byte [] control, AudioFormat format,
    double time)
{
    // Check if the initial delays m_delayStart should be recomputed.  This should
    // only happen if the user changes the number of voices or minimum and maximum
    // delay values in the graphical user interface
    if(! m_delaySet)
    {
        computeDelayStart();
        m_curByteSet = false;
    }

    // Check if the rates of delay variation should be recomputed.  This should
    // only happen if the user changes the rate in the graphical user interface
    if(!m_ratesSet)
        computeRates();

    // Initialize variables
    int blockAlign = (format.getChannels() * format.getSampleSizeInBits()) / 8;
    int sTo = 0;
    int firstUsedBuffer = 0;

    // Store audio data for later use
    byte [] storeBuffer = new byte [dry.length];
    System.arraycopy(dry, 0, storeBuffer, 0, dry.length);
    m_storeBuffers.add(storeBuffer);

    // Compute how far in the past each repetition should look to obtain data.
```

```
// This should only happen on the first call to the apply function, unless
// there were changes by the user
for(int vox = 0; !m_curByteSet && vox < m_vox - 1; vox++)
{
   m_accumulateCurByte[vox] = 0;
   m_curBuffer[vox] = m_storeBuffers.size() - 1;
   m_curByte[vox] = (int) ((double) m_delay[vox] * format.getSampleRate()
      * (double) blockAlign);
   while (m_curByte[vox] > 0 && m_curByte[vox] % blockAlign != 0)
      m_curByte[vox]--;
   m_curByte[vox] = Math.max(m_curByte[vox], 0);
   m_curByte[vox] = -m_curByte[vox];
   while (m_curByte[vox] < 0)
   {
      m_curByte[vox] += dry.length;
      m_curBuffer[vox]--;
   }
}

// The setting of delays is done
m_curByteSet = true;

// Check which buffers are used and which should be removed.  The actual
// removal is below
firstUsedBuffer = m_curBuffer[0];
for(int vox = 1; vox < m_vox - 1; vox++)
   firstUsedBuffer = Math.min(firstUsedBuffer, m_curBuffer[vox]);

byte bufferFrom[] = null;

// Apply the effect
for(int i = 0; i < dry.length; i += blockAlign)
{
   sTo = 0;

   // Create a repetition for each of the chorus voices
   for (int vox = 0; vox < m_vox - 1; vox++)
   {
      if (m_curBuffer[vox] >= 0 && m_curBuffer[vox] < m_storeBuffers.size())
      {
         bufferFrom = m_storeBuffers.get(m_curBuffer[vox]);
         sTo = (int) ((float) sTo + (float) (short)
            (((bufferFrom[m_curByte[vox] + 1] & 0xff) << 8)
            + (bufferFrom[m_curByte[vox]] & 0xff)));
      }

      // Vary the delay of each voice
```

```java
            m_delay[vox] -= m_rates[vox] / format.getSampleRate();

            // Check if we need to change the direction of the delay variation
            if (m_delay[vox] >= m_max || m_delay[vox] <= m_min)
               m_rates[vox] = -m_rates[vox];

            // Check if we need to skip a byte or repeat a byte.  We divide by 1000
            // as the rate of change is in milliseconds per second
            m_accumulateCurByte[vox] += (m_rates[vox] * blockAlign) / 1000D;

            // Skip a byte
            if (m_accumulateCurByte[vox] >= blockAlign)
            {
               m_curByte[vox] += blockAlign;
               m_accumulateCurByte[vox] -= blockAlign;
            }

            // Repeat a byte
            if (m_accumulateCurByte[vox] <= -blockAlign)
            {
               m_curByte[vox] -= blockAlign;
               m_accumulateCurByte[vox] += blockAlign;
            }

            // Increment bytes either way and move to the next stored buffer if
            // necessary
            m_curByte[vox] += blockAlign;
            if (m_curByte[vox] >= dry.length)
            {
               m_curBuffer[vox]++;
               m_curByte[vox] = 0;
            }
         }

         // Store the output
         sTo = Math.min(Math.max(sTo, Short.MIN_VALUE), Short.MAX_VALUE);
         wet[i] = (byte) (sTo & 0xff);
         wet[i + 1] = (byte) (sTo >>> 8 & 0xff);
      }

      // Remove unused stored buffers
      while(firstUsedBuffer > 0)
      {
         m_storeBuffers.remove(0);
         firstUsedBuffer--;
         for(int vox = 0; vox < MAXVOX; vox++)
            m_curBuffer[vox]--;
      }
```

119

```
}
```

Chapter 10. Bass chorus

A *bass chorus* is a chorus, but rather than creating repetitions of the original signal, it creates repetitions of the original signal less its bass frequencies. Repeating the bass frequencies may make the sound muddy and it may be better to take these out before creating the chorus voices.

To remove the bass frequencies from the original signal, if there are any, we use a *high pass filter*. High pass filters first appear in chapter 9 of volume 1.

10.1. High pass filter

Below is the computation of a finite impulse response high pass filter.

Code 53. High pass filter

```
public static void computeHighPassFilter(float [] a, float fc, float fs)
{
    // This function creates a finite impulse response high pass filter.  The
    // filter coefficients are placed in the floating-point array a.  The length
    // of the filter is the length of a.  fc is the cutoff frequency of the
    // filter.  fs is the sampling frequency

    // Create a low pass filter
    for(int i = 0; i < a.length; i++)
    {
        if(i == (a.length - 1) / 2)
            a[i] = 2 * fc / fs;
        else
            a[i] = (float) (Math.sin(2D * Math.PI * fc * (i - (a.length - 1) / 2)
                / fs) / (Math.PI * (i - (a.length - 1) / 2)));
    }

    // Invert the low pass filter into a high pass filter
    for(int i = 0; i < a.length; i++)
        a[i] *= -1F;
    a[(a.length - 1) / 2]++;
}
```

The length of the filter **a** can be set to a fixed value or computed based on the cutoff frequency and some user defined precision (e.g., width of the filter transition band). It should be large enough to accomplish the removal of low frequencies from the signal. For example, the following figure shows the magnitude responses of three high pass filters with different length with cutoff frequency 300 Hz, given the sampling frequency of 44100 Hz.

Figure 5. Magnitude response of high pass filters with three different lengths

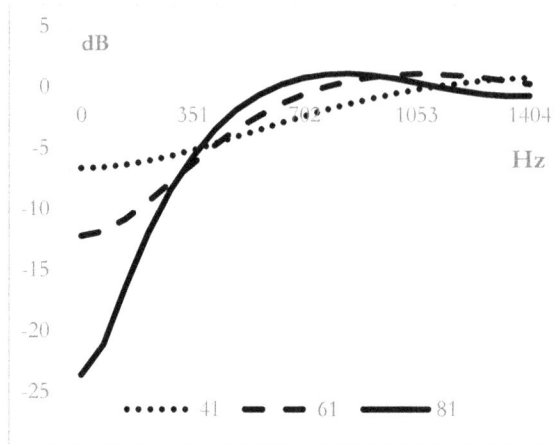

When the sampling frequency is 44100 Hz and the cutoff frequency is 300 Hz, the magnitude response of the high pass filter seems appropriate if the filter is at least of length 81 points.

At this cutoff frequency and sampling rate, the magnitude response of a filter of about 81 points begins to resemble the magnitude response of a high pass filter.

Choosing a filter of 81 points means that the bass chorus will perform at least 81 multiplications and additions for each sample of audio data. This finite impulse response filter is quite computationally intensive. A contemporary computer will handle these computations easily, but we can reduce the number of computations if we use an infinite impulse response filter. We use infinite impulse response filters later in the book, in chapter 15. For now, finite impulse response filters are easier to create and combine and we can control their phase response.

10.2. Implementation of the bass chorus

The member data of the bass chorus are the same as the member data of the chorus with the following additions.

int FILTERLENGTH – This is the length of the high pass filter. It is set to 81.

float m_bassFrequency – This is the cutoff frequency for the high pass filter chosen by the user.

boolean m_filterSet – This is a flag that shows whether the filter coefficients have been computed (**true**) or should be computed (**false**).

float m_h – These are the coefficients of the high pass filter.

Vector<float []> m_storeWetBuffers – Similarly to the chorus effect in the previous chapter, the bass chorus uses **m_storeBuffers** to store the original signal. These data are filtered through a high pass filter. The output of the high pass filter is stored in **m_storeWetBuffers** and is then used by the chorus part of the bass chorus. Unlike **m_storeBuffers**, which stores byte arrays,

m_storeWetBuffers stores floating-point arrays. It makes sense to keep floating-point data for the computation of the chorus part, until these data are stored as bytes in the final output of the bass chorus effect.

The constructor for the bass chorus is identical to the constructor for the chorus, with the following additions.

Code 54. Bass chorus constructor

```
...


// The bass frequency cutoff can also be set by the user
m_bassFrequency = 300F;

// These are the coefficients of the high pass filter.  They are computed
// during playback in case the user modifies them during playback (e.g., if
// the user has access to a precision control that changes the length of the
// high pass filter)
m_h = null;


...
```

In addition, note the change in the function **hasData**. **m_storeWetBuffers** contains the audio data after the high pass filter and is the last point of audio data storage

Code 55. Bass chorus hasData

```
public boolean hasData()
{
   if (m_storeWetBuffers.size() > 0)
      return true;
   return false;
}
```

The following code is executed at the beginning of playback.

Code 56. Bass chorus at the beginning of playback

```
public void startPlay()
{
   // The difference between this function and the same function in the chorus
   // in the previous chapter is the initialization of m_storeWetBuffers
   m_storeWetBuffers.setSize(0);

   m_storeBuffers.setSize(0);
   if(!m_delaySet)
      computeDelayStart();
   if(!m_ratesSet)
      computeRates();
   m_curByteSet = false;
```

```
}
```

The following is the implementation of the bass chorus.

Code 57. Bass chorus

```java
// Compute the high pass filter
public void filterDesign(AudioFormat format)
{
    // Allocate space only if needed
    if (m_h == null)
        m_h = new float[FILTERLENGTH];
    if (m_h.length != FILTERLENGTH)
        m_h = new float[FILTERLENGTH];

    // Calculate the filter coefficients
    computeHighPassFilter(m_h, m_bassFrequency, format.getSampleRate());
    m_filterSet = true;
}


// Apply the effect
public void apply(byte [] dry, byte [] wet, byte [] control, AudioFormat format,
    double time)
{
    // Calculate the initial delays if needed
    if(! m_delaySet)
    {
        computeDelayStart();
        m_curByteSet = false;
    }

    // Calculate the rates of variation in the delays if needed
    if(!m_ratesSet)
        computeRates();

    // Compute the filter if needed
    if(!m_filterSet)
        filterDesign(format);

    // Initialize variables
    int blockAlign = (format.getChannels() * format.getSampleSizeInBits()) / 8;
    int channels = format.getChannels();
    float sTo = 0.0F;
    float sFrom = 0.0F;

    // Store the incoming original audio data
    byte [] storeBuffer = new byte [dry.length];
    System.arraycopy(dry, 0, storeBuffer, 0, dry.length);
```

```
   m_storeBuffers.add(storeBuffer);

   // Allocate space for the audio data after the high pass filter.  Since this
   // is an array of floating points and not bytes, the number of floating
   // points needed is smaller than the number of bytes in the incoming buffers
   float wetf[] = new float[(dry.length * channels) / blockAlign];

   // Use the same structure that we use for delays to calculate how far back
   // the high pass filter should look when processing data.  Each audio sample
   // after the filter should be computed from the previous FILTERLENGTH samples
   // of original data
   int curBufferStart = m_storeBuffers.size() - 1;
   int curByteStart = -FILTERLENGTH * blockAlign;
   while (curByteStart < 0)
   {
      curByteStart += dry.length;
      curBufferStart--;
   }
   byte bufferFromStart[] = null;
   if(curBufferStart >= 0)
      bufferFromStart = m_storeBuffers.get(curBufferStart);

   // Remove unused buffers with original audio data
   while (curBufferStart > 0)
   {
      m_storeBuffers.remove(0);
      curBufferStart--;
   }

   // Apply the high pass filter for each sample of original data
   for(int i = 0, weti = 0; i < dry.length; i += blockAlign,
      curByteStart += blockAlign, weti += channels)
   {
      // Increment the index of the stored original audio buffers if needed
      if(curByteStart >= dry.length)
      {
         curBufferStart++;
         bufferFromStart = m_storeBuffers.get(curBufferStart);
         curByteStart = 0;
      }

      // Initialize data
      int curBuffer = curBufferStart;
      int curByte = curByteStart;
      byte bufferFrom[] = bufferFromStart;
      sTo = 0.0F;

      // Calculate the output of the filter
```

```
    for(int j = 0; j < FILTERLENGTH; j++, curByte += blockAlign)
    {
        // Increment buffer indices if needed
        if(curByte >= dry.length)
        {
            curBuffer++;
            bufferFrom = m_storeBuffers.get(curBuffer);
            curByte = 0;
        }
        if(curBufferStart >= 0)
        {
            sFrom = (short)(((bufferFrom[curByte + 1] & 0xff) << 8)
                + (bufferFrom[curByte] & 0xff));

            // Calculate the output of the high pass filter
            sTo += sFrom * m_h[j];
        }
    }

    // Record the output of the high pass filter
    sTo = Math.max(Math.min(sTo, Short.MAX_VALUE), Short.MIN_VALUE);
    wetf[weti] = sTo;
}

// Store the output of the high pass filter so that it can be used by the
// chorus
m_storeWetBuffers.add(wetf);

int firstUsedBuffer = 0;

// Calculate how far back the chorus delays should look when using stored
// high pass filter output.  Note that blockAlign is not used in here, as
// the stored high pass filter output contains arrays of floating points
// and not bytes
for(int vox = 0; !m_curByteSet && vox < m_vox - 1; vox++)
{
    m_accumulateCurByte[vox] = 0;
    m_curBuffer[vox] = m_storeWetBuffers.size() - 1;
    m_curByte[vox] = (int) ((double) m_delay[vox] * format.getSampleRate()
        * channels);
    while (m_curByte[vox] > 0 && m_curByte[vox] % channels != 0)
        m_curByte[vox]--;
    m_curByte[vox] = Math.max(m_curByte[vox], 0);
    m_curByte[vox] = -m_curByte[vox];
    while (m_curByte[vox] < 0)
    {
        m_curByte[vox] += wetf.length;
```

```
            m_curBuffer[vox]--;
        }
    }

    m_curByteSet = true;

    // Keep track of which buffers are used
    for(int vox = 0; vox < m_vox - 1; vox++)
    {
        if (vox == 0)
            firstUsedBuffer = m_curBuffer[vox];
        else
            firstUsedBuffer = Math.min(firstUsedBuffer, m_curBuffer[vox]);
    }

    float bufferFrom[] = null;

    // Apply the chorus.  This code is identical to the one used for the chorus
    // effect, except it uses stored high pass filter output, rather than the
    // original incoming audio data
    for(int i = 0; i < dry.length; i += blockAlign)
    {
        sTo = 0;

        for (int vox = 0; vox < m_vox - 1; vox++)
        {
            if (m_curBuffer[vox] >= 0
                && m_curBuffer[vox] < m_storeWetBuffers.size())
            {
                bufferFrom = m_storeWetBuffers.get(m_curBuffer[vox]);
                if (m_curByte[vox] >= bufferFrom.length)
                {
                    m_curBuffer[vox]++;
                    bufferFrom = m_storeWetBuffers.get(m_curBuffer[vox]);
                    m_curByte[vox] = 0;
                }
                sTo += (float) bufferFrom[m_curByte[vox]];
            }

            // As in the chorus, vary the delays with time.  Use channels here and
            // not blockAlign as in the chorus, as the output of the high pass
            // filter is stored in an array of floating-point numbers and not bytes
            m_delay[vox] -= m_rates[vox] / format.getSampleRate();
            if (m_delay[vox] >= m_max || m_delay[vox] <= m_min)
                m_rates[vox] = -m_rates[vox];
            m_accumulateCurByte[vox] += (m_rates[vox] * channels) / 1000D;
            if (m_accumulateCurByte[vox] >= channels)
            {
```

```
                m_curByte[vox] += channels;
                m_accumulateCurByte[vox] -= channels;
            }
            if (m_accumulateCurByte[vox] <= -channels)
            {
                m_curByte[vox] -= channels;
                m_accumulateCurByte[vox] += channels;
            }
            m_curByte[vox] += channels;
            if (m_curByte[vox] >= wetf.length)
            {
                m_curBuffer[vox]++;
                m_curByte[vox] = 0;
            }
        }

        // Store the output
        sTo = Math.min(Math.max(sTo, Short.MIN_VALUE), Short.MAX_VALUE);
        wet[i] = (byte) ((int) sTo & 0xff);
        wet[i + 1] = (byte) ((int) sTo >>> 8 & 0xff);
    }

    // Remove stored high pass filter output that is not used
    while(firstUsedBuffer > 0)
    {
        m_storeWetBuffers.remove(0);
        firstUsedBuffer--;
        for (int vox = 0; vox < VOXMAX; vox++)
            m_curBuffer[vox]--;
    }
}
```

10.3. Delays in phase

A high pass filter of length 81 points will delay the bass chorus repetitions by $(81 - 1) / 2 = 40$ samples or $40 / 44100 \approx 0.001$ seconds. If, for example, a repetition should be delayed by 20 ms, it will be delayed by 21 ms because of the high pass filter. We can ignore this difference, as it is small.

If we wanted to adjust for the delay in phase, we can do so. The following line is a part of the code snipped that computes how far back each repetition should look to obtain its audio data.

Code 58. Bass chorus delays unadjusted for the phase delay

```
m_curByte[vox] = (int) ((double) m_delay[vox] * format.getSampleRate()
    * channels);
```

The delay in this line can be adjusted as follows.

Code 59. Bass chorus delays adjusted for the phase delay

```
m_curByte[vox] = (int) ((double) m_delay[vox] * format.getSampleRate()
    * channels) - (FILTERLENGTH - 1) / 2;
```

Chapter 11. Equalizer

An *equalizer* changes the magnitude of frequencies in selected bands, while leaving other frequency bands unchanged. It can be implemented with a single frequency filter, similarly to the high pass filter employed in the bass chorus. Calculating the coefficients of the equalizer filter is only slightly more complex.

A simple equalizer is presented in chapter 11 in volume 1. The equalizer frequency filter is the sum of several separate filters. Each of these filters passes the frequencies in a specific frequency band and stops the frequencies outside of that band. Applying gain to these individual filters means applying gain to the frequencies in the corresponding frequency band, but not to any other frequencies. The output of the equalizer filter is equivalent to the sum of the output of individual filters. With the additional gain provided by the user, the equalizer increases or decreases the gain of frequencies in specific frequency bands.

11.1. Low pass, high pass, and band pass filters

The following are filter computations used in the equalizer. We reexamine the computations for low pass and high pass filters and present computations for band pass and band stop filters.

Code 60. Low pass filter

```
public static void computeLowPassFilter(float [] a, float fc, float fs)
{
    // This function creates a finite impulse response low pass filter.  The
    // filter coefficients are placed in the floating-point array a.  The length
    // of the filter is the length of a.  fc is the cutoff frequency of the
    // filter.  fs is the sampling frequency

    // Calculate the coefficients of the low pass filter
    for(int i = 0; i < a.length; i++)
    {
        if(i == (a.length - 1) / 2)
            a[i] = 2 * fc / fs;
        else
            a[i] = (float) (Math.sin(2D * Math.PI * fc * (i - (a.length - 1) / 2)
                / fs) / (Math.PI * (i - (a.length - 1) / 2)));
    }

    // Apply a Blackman window to the low pass filter to reduce its Gibbs
    // phenomenon ripples
    blackmanWindow(a);
}
```

The difference between this computation and the computation used in creating the bass chorus high pass filter is the Blackman window. We window the low pass filter to reduce the Gibbs phenomenon ripples characteristic of finite impulse response filters. Although adding a window is not required, it makes the magnitude response of the equalizer smoother.

Code 61. Blackman window

```java
public static void blackmanWindow(float [] a)
{
    // This function applies a Blackman window to the filter a.  The length of the
    // window is the same as the length of a.  The coefficients of the filter a
    // must be computed beforehand

    for(int i = 0; i < a.length; i++)
       a[i] *= (0.42 - 0.5D * Math.cos((2D * Math.PI * i)
          / (double) (a.length - 1))) + 0.08 * Math.cos((4D * Math.PI * i)
          / (double) (a.length - 1));
}
```

There are many possible windows and the choice of window depends on the task at hand. For example, if we want a very smooth magnitude response of the equalizer, we might want to add a Bartlett-Hann window. We can do so with the understanding that the magnitude response of a filter with the Bartlett-Hann window may have large transition bands and we therefore cannot create an equalizer that produces sharp changes from the gain in one frequency band to the gain in a neighboring frequency band. The Bartlett-Hann window can be computed as follows.

Code 62. Bartlett-Hann window

```java
public static void bartlettHannWindow(float [] a)
{
    // This function applies a Blackman window to the filter a.  The length of the
    // window is the same as the length of a.  The coefficients of the filter a
    // must be computed beforehand

    for(int i = 0; i < a.length; i++)
       a[i] *= 0.62 - 0.48 * Math.abs((double) i
          / (a.length - 1)) - 0.38 * Math.cos((2D * Math.PI * i)
          / (double) (a.length - 1));
}
```

A high pass filter can be computed from the low pass filter as before.

Code 63. High pass filter

```java
public static void computeHighPassFilter(float [] a, float fc, float fs)
{
    // This function creates a finite impulse response high pass filter.  The
    // filter coefficients are placed in the floating-point array a.  The length
    // of the filter is the length of a.  fc is the cutoff frequency of the
    // filter.  fs is the sampling frequency
```

```
   // Create a low pass filter.  Note that we are inverting a low pass filter
   // with a window and so the high pass filter computed here is also windowed
   computeLowPassFilter(a, fc, fs);

   // Invert the low pass filter into a high pass filter
   for(int i = 0; i < a.length; i++)
      a[i] *= -1F;
   a[(a.length - 1) / 2]++;
}
```

A band stop filter is the sum of a low pass filter and a high pass filter. Here, of course, we expect that if **fclow** is the lower cutoff frequency of the band stop filter and **fchigh** is the higher cutoff frequency, then **fchigh** > **fclow**.

Code 64. Band stop filter

```
public static void computeBandStopFilter(float [] a, float fclow, float fchigh,
   float fs)
{
   // This function creates a finite impulse response band stop filter.  The
   // filter coefficients are placed in the floating-point array a.  The length
   // of the filter is the length of a.  fclow is the bottom cutoff frequency of
   // the filter.  fchigh is the top cutoff frequency of the filter. fs is the
   // sampling frequency

   // Compute a low pass filter at the lower cutoff frequency
   computeLowPassFilter(a, fclow, fs);

   // Compute a high pass filter at the higher cutoff frequency
   float [] b = new float [a.length];
   computeHighPassFilter(b, fchigh, fs);

   // Sum the two filters
   for (int i = 0; i < a.length; i++)
      a[i] = a[i] + b[i];
}
```

A band pass filter is an inverted band stop filter.

Code 65. Band pass filter

```
public static void computeBandPassFilter(float [] a, float fclow, float fchigh,
   float fs)
{
   // This function creates a finite impulse response band pass filter.  The
   // filter coefficients are placed in the floating-point array a.  The length
   // of the filter is the length of a.  fclow is the bottom cutoff frequency of
   // the filter.  fchigh is the top cutoff frequency of the filter. fs is the
   // sampling frequency
```

```
    // Create a band stop filter
    computeBandStopFilter(a, fclow, fchigh, fs);

    // Invert the band stop filter into a band pass filter
    for(int i = 0; i < a.length; i++)
        a[i] *= -1F;
    a[(a.length - 1) / 2]++;
}
```

11.2. The equalizing filter

Conceptually, the equalizer presented here is a collection of low pass, high pass, and band pass filters. These filters split the incoming signal into several signals, each fitting into a separate frequency interval. Thus, some filters split out the low frequencies, some filters split out the mid frequencies, and some filters split the high frequencies. We can use as many such frequency bands as we wish. Once the signals are separated, the equalizer applies some gain to each of them (e.g., to lower mids and boost highs) and puts them back together.

In practice, all equalizer computations are linear, since we are using finite impulse response filters. We can change the order of tasks in the pseudo algorithm in the previous paragraph. Instead of splitting the signal, applying gain, and putting the signal back together, we take the set of equalizer filters, apply gain to them, and combine them into one single equalization filter. We then simply apply this one filter to the signal.

The equalizer presented below is simple. Frequency bands are fixed and cannot be adjusted by the user. The filter length, which determines the precision of the equalizer, is also fixed and cannot be changed by the user. The user can only change the gains applied to each of the frequency bands in the equalizer.

In practice, an equalizer may allow the user to adjust not only the gains on frequency bands, but also the bands themselves and the length of the filter. The structure of the equalizer below can be adjusted easily to allow for such changes. For example, a fixed band and fixed precision equalizer can compute the individual frequency band filters once, in its constructor. Below, these computations are performed during the call to the **apply** function. The equalizer can be changed easily to allow changes to bands and precision. Where such changes should occur is noted below.

The following are the member data of the equalizer.

int NUMBANDS – This is the number of bands in the equalizer. In a fixed band equalizer, this variable is fixed. If the user can change the number of bands, the user should also be allowed to change the actual bands **BANDS** below.

int FILTERLENGTH – This is the length of the filter. The code below works whether this is a final constant or a variable that the user can change. Modifying this variable changes the precision of the equalizer.

float [] BANDS – These are the bands of the equalizer. Suppose, for example, that we have a 10-band equalizer (i.e., **NUMBANDS** = 10). We wish to split the frequency spectrum between 20 Hz and 22000 Hz – the lower and upper limits of the human hearing – into 10 bands. We want to do so evenly, but exponentially, as this is how the human ear perceives frequencies. We set the top of the first band at $(22000 / 20)^{1/10} = 40$ Hz, the top of the second band at $(22000 / 20)^{2/10}$ = 81 Hz and so on. The length of **BANDS** is **NUMBANDS**. Both variables can be final or subject to user changes.

boolean m_filtersSet – This variable shows whether the filter for each of the frequency bands has been computed (**true**) or needs to be computed (**false**). In a fixed band equalizer, the filters for each frequency band can be computed only once, in the equalizer constructor. If the user changes the equalizer bands, this variable should be set to **false** so that the equalizer can recompute the individual band filters accordingly.

boolean m_totalFilterSet – This variable shows whether the equalizer filter has been computed (**true**) or needs to be recomputed (**false**). Even if the individual band filters stay the same, the total equalizer filter must be recomputed any time the user changes the gains applied to the frequency bands.

float [] m_h – This array contains the equalizer filter coefficients. This array can be allocated once in a fixed band equalizer, but must be allocated again, if the user changes **FILTERLENGTH**.

Band [] m_bands – These are the bands of the equalizer. Band is the following simple class.

Code 66. An equalizer band

```
public class Band
{
    public float m_lowCutoff;      // the bottom cutoff frequency
    public float m_highCutoff;     // the top cutoff frequency
    public float m_gain;           // the gain applied to the band
    public float m_h[];            // the band pass filter
}
```

The members of **Band** are as follows.

float m_lowCutoff – This is the bottom end of the frequency band and the bottom cutoff frequency for the corresponding filter.

float m_highCutoff – This is the top end of the frequency band and the top cutoff frequency of the corresponding filter.

float m_gain – This is the gain that the user applies to the signal in the frequency band. The gain is measured in decibels.

float [] m_h – This is the low pass, high pass, or band pass filter that is used to "extract" the frequencies in the frequency band.

The equalizer constructor sets up the frequency bands as follows.

Code 67. Equalizer bands

```
m_bands = new Band[NUMBANDS];
for(int i = 0; i < NUMBANDS; i++)
{
   m_bands[i].m_h = null;
   m_bands[i].m_gain = 0;
}
m_bands[0].m_lowCutoff = 0;
m_bands[0].m_highCutoff = BANDS[0];
for(int i = 1; i < NUMBANDS; i++)
{
   m_bands[i].m_lowCutoff = BANDS[i - 1];
   m_bands[i].m_highCutoff = BANDS[i];
}
```

The individual band filters can be computed as follows.

Code 68. Equalizer band filters

```
public void computeFilter(AudioFormat format)
{
   // The audio format of incoming data is needed, as the filter calculations
   // use the sampling rate

   // Allocate the filter arrays
   for(int i = 0; i < NUMBANDS; i++)
   {
      if (m_bands[i].m_h == null)
         m_bands[i].m_h = new float[FILTERLENGTH];
      if (m_bands[i].m_h.length != FILTERLENGTH)
         m_bands[i].m_h = new float[FILTERLENGTH];
   }

   // The first (lowest) band uses a low pass filter
   computeLowPassFilter(m_bands[0].m_h, m_bands[0].m_highCutoff,
      format.getSampleRate());

   // The last (highest) band uses a high pass filter
   computeHighPassFilter(m_bands[NUMBANDS - 1].m_h,
      m_bands[NUMBANDS - 1].m_lowCutoff, format.getSampleRate());

   // The remaining bands use band pass filters
   for(int i = 1; i < NUMBANDS - 1; i++)
```

```
{
    computeBandPassFilter(m_bands[i].m_h, m_bands[i].m_lowCutoff,
        m_bands[i].m_highCutoff, format.getSampleRate());
}

m_filtersSet = true;
}
```

The total equalizer filter is simply the sum of the individual band filters, with the corresponding gain applied.

Code 69. Total equalizer filter

```
public void computeTotalFilter()
{
    // allocate and initialize the total filter array
    if(m_h == null)
        m_h = new float[FILTERLENGTH];
    else if(m_h.length != FILTERLENGTH)
        m_h = new float[FILTERLENGTH];
    for(int i = 0; i < FILTERLENGTH; i++)
        m_h[i] = 0.0F;

    // Simply sum the individual band filters with the appropriate gain
    // the gain is assumed to be in decibels and must be converted to a multiple
    for(int i = 0; i < NUMBANDS; i++)
    {
        float gain = (float) Math.pow(10D, m_bands[i].m_gain / 20F);
        for(int j = 0; j < FILTERLENGTH; j++)
            m_h[j] += gain * m_bands[i].m_h[j];
    }

    m_totalFilterSet = true;
}
```

11.3. Implementation of the equalizer

The following is the constructor of the equalizer.

Code 70. Equalizer constructor

```
public Equalizer()
{
    // Create the equalizer bands based on the information in BANDS
    m_bands = new Band[NUMBANDS];
    for(int i = 0; i < NUMBANDS; i++)
    {
        m_bands[i].m_h = null;
        m_bands[i].m_gain = 0;
```

```
   }
   m_bands[0].m_lowCutoff = 0;
   m_bands[0].m_highCutoff = BANDS[0];
   for(int i = 1; i < NUMBANDS; i++)
   {
      m_bands[i].m_lowCutoff = BANDS[i - 1];
      m_bands[i].m_highCutoff = BANDS[i];
   }

   // The total equalizer filter is computed during playback, in case the
   // user makes changes to the gains of individual bands (or to the bands
   // themselves, if this is a parametric equalizer)
   m_h = null;
   m_filtersSet = false;
   m_totalFilterSet = false;

   m_storeBuffers = new Vector<byte []>(0);
}
```

The following code is executed at the beginning of playback.

Code 71. Equalizer at the beginning of playback

```
public void startPlay()
{
   m_filtersSet = false;
   m_totalFilterSet = false;
   m_storeBuffers.setSize(0);
}
```

Once the filter is computed, the implementation of the equalizer is the same as the application of the high pass filter in the bass chorus.

Code 72. Equalizer

```
public void apply(byte [] dry, byte [] wet, byte [] control, AudioFormat format,
   double time)
{
   // Calculate the individual filters, if needed.  This must happen if the user
   // changes the filter length (precision) or the position of the frequency
   // bands
   if(!m_filtersSet)
      computeFilter(format);

   // Calculate the total equalizer filter, if needed.  This must happen if the
   // user changes the filter length, the frequency bands, or the gain applied to
   // any of the bands
   if(!m_totalFilterSet)
      computeTotalFilter();
```

```
// Initialize variables
int blockAlign = (format.getChannels() * format.getSampleSizeInBits()) / 8;
float sFrom = 0.0F;
float sTo = 0.0F;

// Store incoming audio data
byte [] storeBuffer = new byte [dry.length];
System.arraycopy(dry, 0, storeBuffer, 0, dry.length);
m_storeBuffers.add(storeBuffer);

// Calculate how far back the equalizer should look to apply the filter.  At
// each sample, the filter uses the previous FILTERLENGTH samples
int curBufferStart = m_storeBuffers.size() - 1;
int curByteStart = -FILTERLENGTH * blockAlign;
while (curByteStart < 0)
{
    curByteStart += dry.length;
    curBufferStart--;
}
byte bufferFromStart[] = null;
if(curBufferStart >= 0)
    bufferFromStart = m_storeBuffers.get(curBufferStart);

// Remove unused past audio buffers
while (curBufferStart > 0)
{
    m_storeBuffers.remove(0);
    curBufferStart--;
}

// Apply the filter at each sample
for(int i = 0; i < dry.length; i += blockAlign, curByteStart += blockAlign)
{
    if(curByteStart >= dry.length)
    {
        curBufferStart++;
        bufferFromStart = m_storeBuffers.get(curBufferStart);
        curByteStart = 0;
    }

    int curBuffer = curBufferStart;
    int curByte = curByteStart;
    byte bufferFrom[] = bufferFromStart;
    sTo = 0.0F;

    // Apply the filter at a sample
```

```
        for(int j = 0; j < FILTERLENGTH; j++, curByte += blockAlign)
        {
            if(curByte >= dry.length)
            {
                curBuffer++;
                bufferFrom = m_storeBuffers.get(curBuffer);
                curByte = 0;
            }
            if(curBufferStart >= 0)
            {
                sFrom = (short)(((bufferFrom[curByte + 1] & 0xff) << 8)
                    + (bufferFrom[curByte] & 0xff));
                sTo += sFrom * m_h[j];
            }
        }

        // Store the output
        sTo = Math.max(Math.min(sTo, Short.MAX_VALUE), Short.MIN_VALUE);
        wet[i] = (byte)((int)sTo & 0xff);
        wet[i + 1] = (byte)((int)sTo >>> 8 & 0xff);
    }
}
```

Note that the equalizer uses filters of equal length and, for neighboring filters, with the same cutoff frequencies. This ensures that a smooth transition from one frequency band to another. If, for example, the user specifies 0 dB gain in each of the frequency bands, this equalizer should produce a flat magnitude response (i.e., without any peaks, dips, or ripples).

11.4. Magnitude response of the equalizer

Suppose that we want to calculate how the filter affects frequencies in the frequency spectrum. The actual magnitude response of the equalizer is different than the desired one. Perhaps we want to draw this actual response for the user.

The following function calculates the magnitude response of the filter for a frequency.

Code 73. Magnitude response of the equalizer

```
public float magnitudeResponse(float [] a, float f, float fs)
{
    // a is the filter, the magnitude response of which we want to compute
    // f is the frequency at which the magnitude response is computed
    // fs is the sampling frequency

    // Compute the angular frequency corresponding to the frequency f
    double w = 2 * Math.PI * f / fs;

    // Since our filters are symmetric around their middles, calculate the
    // the middle to reduce the computations that follow
```

```
    int M = (a.length - 1) / 2;

    // Compute the magnitude response at f
    float output = a[M];
    for(int i = 0; i < M; i++)
        output += 2 * a[i] * Math.cos(w * (i - M));

    // The result is the magnitude response of the filter, as a multiple
    // (i.e., a response of 1 means there were no changes to the magnitude
    // of f
    return Math.abs(output);
}
```

11.5. Phase response of the equalizer

Note that the equalizer does not allow dry and wet mix. There should be no reason to mix the original and the equalized signal and hence there is no reason to compute the phase.

Code 74. Dry and wet mix in the equalizer

```
public boolean allowsDryWetMix()
{
    return false;
}
```

If we wanted to allow dry and wet mix, we must recognize that the equalized wet signal is delayed by **(FILTERLENGTH - 1) / 2**, assuming **FILTERLENGTH** is an odd number. This means that we also must delay the original signal by the same number of samples.

Chapter 12. Noise gate

A *noise gate* zeroes out the signal into silence, when the amplitude of the signal falls below a certain threshold, and leaves the signal unchanged, when the amplitude of the signal is above the threshold. This chapter presents a simple noise gate – one that changes the signal immediately as the signal crosses the threshold. The next chapter presents a compressor that similarly adjusts the amplitude of the signal based on a threshold, but does so gradually.

To adjust signal amplitude, a noise gate must know what that amplitude is. Unlike the distortion, the noise gate cannot use the value of the signal at each sample – the signal peak amplitude – but needs a measure of signal power. The noise gate can use a root mean square (RMS) measure of the signal amplitude, but that would be too slow, as it would require the processing of too much data at each sample. The alternative is to use a *Hilbert transform*. Hilbert transforms are discussed in chapter 23 of volume 1.

12.1. Hilbert transform

The Hilbert transform is a finite impulse response filter. Computing its output is the same as computing the output of the high pass filter of the bass chorus. The coefficients of the Hilbert transform filter be computed as follows.

Code 75. Hilbert transform

```
public static void computeHilbertTransform(float [] a)
{
   // This function calculates the coefficients of the Hilbert transform
   // The coefficients are placed in the argument a.  It is expected that
   // a is of even length (the computations for a Hilbert transform of odd length
   // are different)

   // Since the Hilbert transform coefficients alternate between zero and
   // nonzero, prepare the array first
   for(int i = 0; i < a.length; i++)
      a[i] = 0;

   // Calculate the nonzero coefficients
   for(int i = 1; i < a.length; i += 2)
      a[i] = (float) ((2F / a.length)
         * Math.sin(Math.PI * (i - a.length / 2) / 2)
         * Math.sin(Math.PI * (i - a.length / 2) / 2)
         / Math.tan(Math.PI * (i - a.length / 2) / a.length));
}
```

The Hilbert transform should not be too short. As with most discrete audio filters, the Hilbert transform is not precise at all frequencies. It acts as a band pass filter that attenuates the magnitudes of low frequencies, which means that computing the amplitude envelope with the Hilbert transform at very low and very high frequencies does not work as needed.

As with the bass chorus high pass filter, we can experiment with different Hilbert transform lengths. The following figure shows the magnitude response of Hilbert transforms with three different lengths. Note that a Hilbert transform of 100 points at the sampling rate 44100 Hz attenuates most bass frequencies (below 300 Hz) and only the Hilbert transform of 300 points captures most (although not all) bass frequencies. If we want to use the noise gate on a bass, we should deploy a Hilbert transform of 300 points, if not more.

Figure 6. Magnitude response of the Hilbert transform

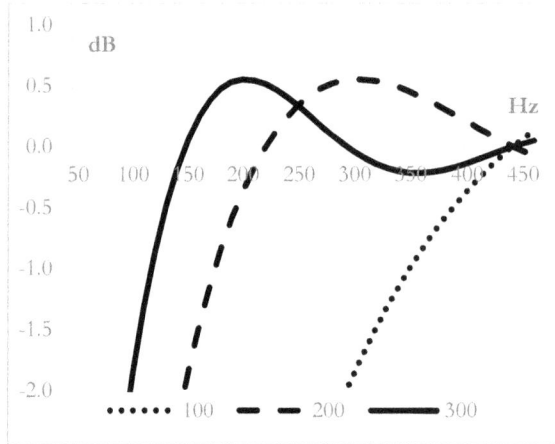

The Hilbert transform acts as a band pass filter and attenuates very low frequencies (as well as very high frequencies). This means that we cannot use the Hilbert transform to calculate the amplitude of low frequencies and therefore we cannot have a noise gate that works on low frequencies. A Hilbert transform of 100 points at the sampling frequency 44100 Hz, for example, attenuates all frequencies below approximately 400 Hz. A Hilbert transform of 200 points attenuates frequencies below 200 Hz. A Hilbert transform of 300 points attenuates frequencies below 150 Hz.

12.2. Implementation of the noise gate

The following are the member data of the noise gate.

int FILTERLENGTH – This is the size of the Hilbert transform. It is set to 300.

float m_threshold – The amplitude threshold of the compressor. It is measured in decibels and can be between, say, -50 dB and 0 dB, where 0 dB is the maximum possible amplitude for the signal. The noise gate begins to act when the amplitude of the signal falls below this threshold. The noise gate leaves the signal unchanged if the amplitude of the signal is above this threshold.

float [] m_hilbert – These are the coefficients of the Hilbert transform.

boolean m_filterSet – This variable shows whether the Hilbert transform coefficients have been computed (**true**) or should be computed (**false**).

Vector<byte []> m_storeBuffers – As in all other effects, this stores past audio data.

The constructor of the noise gate is as follows.

Code 76. Noise gate constructor

```
public NoiseGate()
{
   // The threshold should be set by the user
   m_threshold = -30;

   // The Hilbert transform coefficients could be computed here.  In a more
   // complex implementation, computing these at the beginning of playback
   // allows us access to the format of audio data and we can adjust the
   // length of the transform based on the sampling rate
   m_hilbert = null;
   m_filtersSet = false;

   m_storeBuffers = new Vector<byte []>(0);
}
```

The following code is executed at the beginning of playback.

Code 77. Noise gate at the beginning of playback

```
public void startPlay()
{
   // Again, we could compute the coefficients of the Hilbert transform here,
   // but we could also do it in the apply function below.  The benefit of
   // doing so in the apply function is that the transform will react to user
   // changes (e.g., to changing the length of the Hilbert transform for
   // precision).  This code does not allow the user to make such changes, but
   // it can be changed to do so
   m_storeBuffers.setSize(0);
}
```

The following is the implementation of the noise gate.

Code 78. Noise gate

```
// Calculate the coefficients of the Hilbert transform
public void computeFilters(AudioFormat format)
{
   if (m_hilbert == null)
      m_hilbert = new float[FILTERLENGTH];
   if (m_hilbert.length != FILTERLENGTH)
      m_hilbert = new float[FILTERLENGTH];
   computeHilbertTransform(m_hilbert);
   m_filtersSet = true;
}

public void apply(byte [] dry, byte [] wet, byte [] control, AudioFormat format,
```

```
   double time)
{
   // Compute the filter if needed
   if(!m_filtersSet)
      computeFilters(format);

   // Initialize variables
   int blockAlign = (format.getChannels() * format.getSampleSizeInBits()) / 8;
   float sFrom = 0.0F;
   float sTo = 0;

   // Store audio data for later use
   byte [] storeBuffer = new byte [dry.length];
   System.arraycopy(dry, 0, storeBuffer, 0, dry.length);
   m_storeBuffers.add(storeBuffer);

   // Compute how far back we should look to apply the Hilbert transform filter
   int curBufferStart = m_storeBuffers.size() - 1;
   int curByteStart = FILTERLENGTH * blockAlign;
   curByteStart = -curByteStart;
   while (curByteStart < 0)
   {
      curByteStart += dry.length;
      curBufferStart--;
   }
   byte bufferFrom [] = null;
   if(curBufferStart >= 0)
      bufferFrom = m_storeBuffers.get(curBufferStart);

   // The amplitude envelope is the sum of the output of the Hilbert transform
   // filter and the original signal.  Since the Hilbert transform filter delays
   // the signal by half of the length of the filter, the original signal must be
   // delayed too.  Compute how far back we should look to get the value of the
   // original signal that should be added to the output of the filter
   int curBufferStartIn = m_storeBuffers.size() - 1;
   int curByteStartIn = (FILTERLENGTH / 2) * blockAlign;
   curByteStartIn = -curByteStartIn;
   while (curByteStartIn < 0)
   {
      curByteStartIn += dry.length;
      curBufferStartIn--;
   }
   byte bufferFromIn [] = null;
   if(curBufferStartIn >= 0)
      bufferFromIn = m_storeBuffers.get(curBufferStartIn);

   // Remove past data buffers that are no longer used
```

```
while (curBufferStart > 0 && curBufferStartIn > 0)
{
    m_storeBuffers.remove(0);
    curBufferStart--;
    curBufferStartIn--;
}

// Apply the noise gate
for(int i = 0; i < dry.length; i += blockAlign)
{
    int curBuffer = curBufferStart;
    int curByte = curByteStart;
    sFrom = 0;

    if (curBuffer >= 0 && curBufferStartIn >= 0)
    {
        // Compute the Hilbert transform
        for(int j = 0; j < FILTERLENGTH; j++, curByte += blockAlign)
        {
            if(curByte >= dry.length)
            {
                curBuffer++;
                if (curBuffer < m_storeBuffers.size())
                    bufferFrom = m_storeBuffers.get(curBuffer);
                curByte = 0;
            }
            sFrom += ((short)(((bufferFrom[curByte + 1] & 0xff) << 8)
                + (bufferFrom[curByte] & 0xff)) ) * m_hilbert[j]
                / (float) Short.MAX_VALUE;
        }

        // Add the delayed original signal to the Hilbert transform to
        // get the amplitude envelope of the signal.  We do not actually need to
        // store the computed audio envelope for more than one sample (i.e., we
        // will not use m_storeWetBuffers as in the bass chorus.  (Also, we use
        // "sTo" just so we do not have to declare another variable, but this is
        // not the output signal.  sTo is the delayed original signal.  We delay
        // this signal by the same amount the Hilbert transform delays its
        // output, so that we can add the two signals and obtain the amplitude
        // envelope)
        sTo = ((short)(((bufferFromIn[curByteStartIn + 1] & 0xff) << 8)
            + (bufferFromIn[curByteStartIn] & 0xff)) ) / (float) Short.MAX_VALUE;
        sFrom = (float) Math.sqrt(sTo * sTo + sFrom * sFrom);

        // Convert the threshold from decibels to a multiple
        float currentValue = -60F;
        if (Math.abs(sFrom) > 0)
            currentValue = (float) (20D* Math.log10(Math.abs(sFrom)));
```

```java
            // If the current amplitude is below the threshold, silence the signal
            if (currentValue < m_threshold)
                sTo = 0;
        }

        // Increment indices for the start of the Hilbert transform computation
        curByteStart += blockAlign;
        if(curByteStart >= dry.length)
        {
            curBufferStart++;
            curByteStart = 0;
            if(curBufferStart < m_storeBuffers.size())
                bufferFrom = m_storeBuffers.get(curBufferStart);
        }

        // Increment indices for the delayed original signal
        curByteStartIn += blockAlign;
        if(curByteStartIn >= dry.length)
        {
            curBufferStartIn++;
            curByteStartIn = 0;
            if(curBufferStartIn < m_storeBuffers.size())
                bufferFromIn = m_storeBuffers.get(curBufferStartIn);
        }

        // Store output
        sTo = Math.max(Math.min(sTo, Short.MAX_VALUE), Short.MIN_VALUE);
        wet[i] = (byte)((int) sTo & 0xff);
        wet[i + 1] = (byte)((int) sTo >>> 8 & 0xff);
    }
}
```

Chapter 13. Compressor

The *compressor* below resembles the noise gate of the previous chapter. It must calculate the amplitude of the signal and adjust it depending on how this amplitude compares to a threshold. Rather than doing so abruptly, the compressor does so gradually.

13.1. Practical compression and expansion of dynamics

Practical compressors have at least a *threshold* and a *compression ratio*. The threshold is the amplitude level, above which the compressor acts to change the amplitude dynamics of the sound, and below which the compressor does nothing. The compression ratio is the ratio by which the amplitude of the sound over the threshold is changed. Set, for example, a threshold of -20 dB and a compression ratio at 4:1 (the threshold usually is measured by the difference between the signal and the maximum signal level that the equipment or software can handle). If the sound level exceeds -20 dB it will be brought down by a ratio of 4 to 1. If, for example, the sound level is at -10 dB, which is 10 dB over the threshold, it will be brought to 2.5 dB over the threshold, to -17.5 dB. In simple compressors, the sound at levels under -20 dB will remain unchanged. The result is that loud parts are brought down closer to soft parts. Simple expansion works similarly, except that louder parts will be made even louder and moved further away from softer parts, which could be useful for bringing different instruments together along the time line.

Compressors usually also have an *attack time* and a *release time*. The attack time is the time it takes for the compressor to bring the gain down after detecting an amplitude above the threshold. If, for example, the attack time is set at 50 milliseconds, the first 50 ms of sound after the signal crosses the threshold will be used to get to the desired compression ratio from having no compression, whether the signal is above or below the threshold. The release time is the time it takes for the compressor to stop working after the sound level drops below the threshold. If, for example, the release time is 500 ms the compressor will change the gain over 500 ms after the sound drops under the threshold until it reaches a point of no compression); attack and release times that are dependent on the level of the input signal.

These are the parameters implemented in the code below. Compressors can be much more complex and we show some extensions after implementing the simple compressor. Complex compressors may have: more than one threshold; separate compression / or expansion ratios above and below the threshold; slow or fast transition from one compression ratio to another (*soft knee* or *hard knee*); more than one threshold; input or output gain that will additionally amplify all sound that comes in or goes out (*compensation gain*); response to instantaneous peaks in amplitude or to average of the input gain over a short time period; compression on all channels whenever one channel needs to be compressed or treating each channel independently; and, changes to one signal based on how another signal compares to the set thresholds (*side chaining*). Side chaining specifically can be used on a signal to a time-displaced version of itself. Typically, the compressor will modify a delayed version of the signal, but will be triggered by a non-delayed version of the same signal. This allows the compressor to act quickly and even preemptively, thus creating much smoother dynamics. Such compressors are said to be *looking ahead* or *forward-looking*.

Compression can be used to control the amplitude dynamics of instruments, as they may have different qualities when recorded loud or soft. The natural qualities (*timbre*) of soft and loud parts can be used without losing the soft parts in the mix or overpowering the mix with the loud parts. Compression can also be used to add sustain to instruments that decay quickly. Compressing cymbals, for example, ensures that the cymbal initial hit on the cymbal is brought closer to the cymbal decay. Adding compensation gain then brings the cymbals up in the mix and adds sustain. Compression is often used in *de-essing*. Sibilance frequencies in vocals can be filtered out from the mix with an equalizer and used to trigger the compressor. The compressor then acts only when those frequencies are present effectively lowering sibilance in the mix. Expansion can be used to bring instruments closer together in the mix. If an expansion is applied to the bass and side chained to the drums, the bass will thump with the drums. Another reason to use compression is that songs that are less dynamic can take less storage space (especially in the digital recording world, where you can work with smaller numbers to carry the information). Radio signals are compressed for that reason. A limiter, which is simply a compressor with a very high compression ratio and with very fast attack and release (usually ratio higher than 10:1, attack of less than 5 ms, and a release of up to 20 ms), can be designed to prevent overload on analogue systems (overloads in digital systems do not occur in the same way, as the amplitude of digital audio is naturally limited by its sampling resolution). A limiter that has a compression ratio of 20:1 or over is known as a *hard limiter* or a *brick wall limiter*.

Compressors and expanders can also be combined. When compressors are combined in parallel, usually one of them uses minimal compression ratios whereas the other uses higher compression ratios. When the two output signals are combined, the gain of signals with smaller gain is boosted, with little change to the dynamics of the signals with higher gain, thus improving the low gain recordings. When compressors are used one after the other, one compressor will have smaller ratios and slower attacks and release to smooth out the signal, whereas the other compressor will have higher ratios and faster attack and release to control significantly higher peaks.

13.2. An example of drum compression

This section shows the effect of a compressor on a drum snare and should help understand what a compressor does. There are two reasons to look at drums. The compressor settings that are usually recommended for getting sharper drums in the mix are all the same. These settings are used below. Also, some drum hits – snares and kicks specifically – are very short. Even with the snare (and drum frame) ring, a recorded snare sample is almost always less than one second long, which makes it easy to work with and display visually.

The following figure shows an example recording of an actual snare hit.

Figure 7. An example snare hit

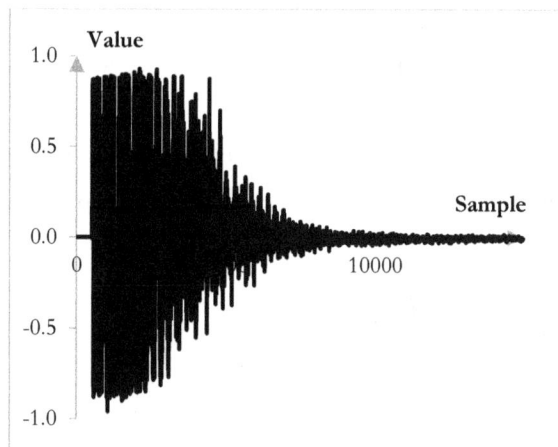

This is an actual 44.1 kHz 16-bit recording of a Yamaha Recording Custom 14x10 inch snare. The digital recorded samples were extracted from the wave file and placed in Microsoft Excel to produce the graph above. The snare was recorded with an SM57 microphone pointed at the top edge of the snare (there was no microphone pointing to the snare bottom).

As can be seen on the graph, the ring of this snare hit is quite long – over 5000 samples, which translates to over 0.1 seconds with the 44.1 kHz sampling rate. We will use a compressor to decrease this ring, while preserving the initial accent, thereby making the snare sharper in the mix.

The following is the amplitude envelope of the snare recording. This is normalized amplitude. In 16-bit recordings, the absolute value of recorded samples can go from 0 to 2^{15}, but we can divide these values by 2^{15} to normalize everything to the range of 0 to 1.

Figure 8. Amplitude envelope of an example snare hit

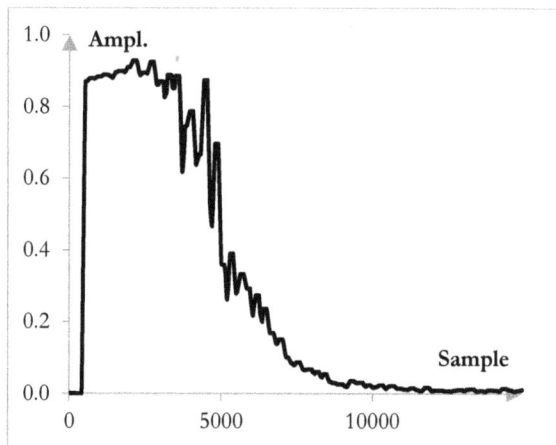

Normalized amplitude of the snare measured as an RMS amplitude. The RMS amplitude is easier to compute than the Hilbert transform for this short sound sample.

The typical compressor settings for getting sharper drums in the mix are: 1) use a threshold of -5 dB; 2) compress amplitudes above the threshold by a ratio of 3 to 1; and 3) use a compressor attack of 15 ms and a compressor release of 50 ms. In the range of 0 to 1, the -5 dB threshold translates to the gain $10^{-5/20} = 0.562$.

The -5 dB compression threshold on the amplitude envelope above is as in the following figure.

Figure 9. Applying compression on the amplitude envelope of the snare hit

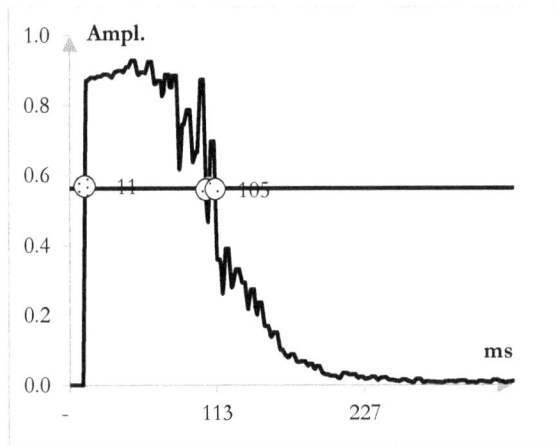

The horizontal line represents the threshold. The signal exceeds the threshold early and falls under the threshold later. Note the change from "samples" to "milliseconds" on the horizontal axis.

At about 11 ms from the beginning of the recording, the amplitude of the snare will exceed the threshold. The compressor will begin decreasing the amplitude of the recorded snare from that point on and, over the next 15 ms, will reach a compression ratio of 3:1. The amplitude of the signal reaches about 0.9, which is about -0.9 dB = $20 \log_{10} 0.9$. This is 4.1 dB above the threshold and will be compressed to 4.1 / 3 = 1.3 dB over the threshold.

Somewhere around 105 ms into the recording, the snare amplitude will drop below the threshold and the compressor will start releasing – changing the compression ratio from 3:1 back to 1:1 (no compression) (if we ignore the quick crossing of the compression threshold back and forth around there). The compressor will reach the 1:1 ratio 50 ms after the release.

The resulting amplitude envelope, compared to the original one, will be the following.

Figure 10. Amplitude envelope of the snare hit before and after the compression

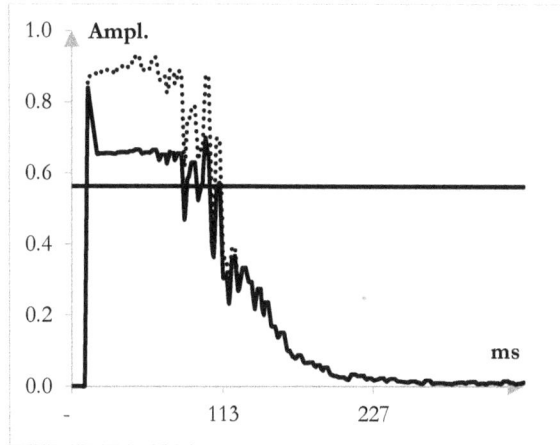

The horizontal line is the threshold. The dotted line is the original amplitude envelope. The solid line is the amplitude envelope after the compression.

Note that it takes some time (the 15 ms attack) for the compressor to drop the amplitude and some time (the 50 ms release) to bring back the amplitude to its original level. The 15 ms attack here is very important. It allows the compressor to preserve the initial hit if the snare. The release is also important. It allows the compressor to continue dropping the ring of the snare smoothly after the amplitude of the signal has dropped below the threshold. Note also that the compressor does not drop the amplitude level so far that it drops below the threshold. It simply changes the amount by which the signal overshoots the threshold.

The original and compressed snare hits are shown below.

Figure 11. Original and compressed snare hit

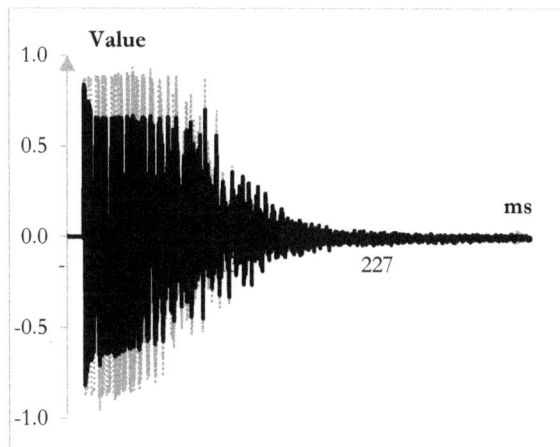

The dotted gray line is the original recording of the snare. The solid black is the compressed snare hit.

Thus, the compressor preserves some of the initial accent and decreases the amplitude of the remaining ring. Since the signal amplitude crosses the threshold a bit before the peak of the amplitude of the initial hit, the initial hit is slightly lower. Some additional output gain (perhaps 1 dB) should be added to the signal. In fact, since now the snare may sound very different, more than 1 dB could be needed.

13.3. Implementation of the simple compressor

The following are the member data of the compressor.

float m_numerator – This is the numerator of the compression ratio. For example, this variable will be equal to 4 in 4:1 compression. The numerator of a typical compressor should be between 1 and 10, but could be larger in limiters.

float m_denominator – This is the denominator of the compression ratio. The denominator is also between 1 and 10. The numerator and denominator cannot both be larger than 1 at the same time. The compressor acts as a compressor if the numerator exceeds 1. The compressor acts as an expander if the denominator exceeds 1. The compressor leaves the signal unchanged if both the numerator and the denominator are 1.

float m_threshold – This is the amplitude threshold of the compressor. It is measured in decibels and can be between, say, -50 dB and 0 dB, where 0 dB is the maximum possible amplitude for the signal. The compressor begins to act when the amplitude of the signal exceeds this threshold. The compressor leaves the signal unchanged if the amplitude of the signal is below this threshold.

float m_attack – This is the attack of the compressor. It is the time from the moment the signal exceeds the threshold to the moment the compressor achieves the compression ratio specified by **m_numerator** and **m_denominator**. It is measured in milliseconds and can be between 0 ms and 200 ms for example.

float m_release – This is the release of the compressor. It is the time from the moment the signal falls below the threshold to the moment the compressor achieves a ratio of no compression and no expansion (i.e., 1:1). The release is measured in milliseconds and can be between 0 ms and 2000 ms for example.

int FILTERLENGTH – This is the size of the Hilbert transform. As with the noise gate, it is set here to 300.

float [] m_hilbert – These are the coefficients of the Hilbert transform.

boolean m_filterSet – This variable shows whether the Hilbert transform coefficients have been computed (**true**) or should be computed (**false**).

Vector<byte []> m_storeBuffers – As in all other effects, this variable stores past audio data.

float m_compressorGain – This is the gain applied by the compressor. We must keep track of it, as the gain applied by the compressor to a specific sample of audio depends not only on the amplitude of the sample itself, but also on the amplitude of previous samples. It is adjusted slowly up and down, depending on the compressor attack and release.

The compressor constructor is as follows.

Code 79. Compressor constructor

```
public Compressor()
{
   // These five items should be set by the user.  If the user does not make
   // changes, the following values specify a compression ratio of 3:1, threshold
   // of -30 dB, attack of 0.25 milliseconds, and release of 50 milliseconds
   m_numerator = 3;
   m_denominator = 1;
   m_threshold = -30;
   m_attack = 0.25F;
   m_release = 50;

   m_hilbert = null;
   m_filtersSet = false;
   m_storeBuffers = new Vector<byte []>(0);
   m_compressorGain = 1;
}
```

The following code is executed at the beginning of playback.

Code 80. Compressor at the beginning of playback

```
public void startPlay()
{
   // Recalculate the Hilbert transform, initialize the compressor gain, and
   // clear past audio data
   m_filtersSet = false;
   m_compressorGain = 1;
   m_storeBuffers.setSize(0);
}
```

The implementation of the compressor is as follows.

Code 81. Compressor

```
public void computeFilters(AudioFormat format)
{
   // This function is identical to the one used for the noise gate.  The Hilbert
   // transform also has the same coefficients as the Hilbert transform for
   // the noise gate
   if (m_hilbert == null)
      m_hilbert = new float[FILTERLENGTH];
   if (m_hilbert.length != FILTERLENGTH)
      m_hilbert = new float[FILTERLENGTH];
   computeHilbertTransform(m_hilbert);
```

```
   m_filtersSet = true;
}

public void apply(byte [] dry, byte [] wet, byte [] control, AudioFormat format,
   double time)
{
   // Recompute the Hilbert transform, if needed.  In this implementation, there
   // is no need to do this here, but other implementations may allow the user
   // to change the length and precision of the Hilbert transform.  In such
   // implementations, it is best to do this here, rather than in the
   // constructor or in startPlay
   if(!m_filtersSet)
      computeFilters(format);

   // Initialize variables
   int blockAlign = (format.getChannels() * format.getSampleSizeInBits()) / 8;
   float sTo = 0;
   float sFrom = 0;

   // Store audio data for later use
   byte [] storeBuffer = new byte [dry.length];
   System.arraycopy(dry, 0, storeBuffer, 0, dry.length);
   m_storeBuffers.add(storeBuffer);

   // In this effect, we keep track of two positions in the audio data.  The
   // first position is how far back we must look to calculate the Hilbert
   // transform, just as with any other filter
   int curBufferStart = m_storeBuffers.size() - 1;
   int curByteStart = -FILTERLENGTH * blockAlign;
   while (curByteStart < 0)
   {
      curByteStart += dry.length;
      curBufferStart--;
   }
   byte bufferFrom [] = null;
   if(curBufferStart >= 0)
      bufferFrom = m_storeBuffers.get(curBufferStart);

   // The second position is kept, because we need to delay the original signal
   // With the Hilbert transform, the amplitude envelope of the signal is
   // computed by adding the output of the transform to the original signal.  In
   // order to do so, the original signal and the Hilbert transform must be
   // delayed by the same number of samples (in this case FILTERLENGTH / 2)
   int curBufferStartIn = m_storeBuffers.size() - 1;
   int curByteStartIn = -(FILTERLENGTH / 2) * blockAlign;
   while (curByteStartIn < 0)
   {
```

```
        curByteStartIn += dry.length;
        curBufferStartIn--;
    }
    byte bufferFromIn [] = null;
    if(curBufferStartIn >= 0)
        bufferFromIn = m_storeBuffers.get(curBufferStartIn);

    // Remove unused buffers
    while (curBufferStart > 0 && curBufferStartIn > 0)
    {
        m_storeBuffers.remove(0);
        curBufferStart--;
        curBufferStartIn--;
    }

    // Apply the effect
    for(int i = 0; i < dry.length; i += blockAlign)
    {
        int curBuffer = curBufferStart;
        int curByte = curByteStart;
        sFrom = 0.0F;

        if (curBuffer >= 0 && curBufferStartIn >= 0)
        {
            // Compute the Hilbert transform to obtain the amplitude of the signal
            for(int j = 0; j < FILTERLENGTH; j++, curByte += blockAlign)
            {
                if(curByte >= dry.length)
                {
                    curBuffer++;
                    if (curBuffer < m_storeBuffers.size())
                        bufferFrom = m_storeBuffers.get(curBuffer);
                    curByte = 0;
                }
                sFrom += ((short)(((bufferFrom[curByte + 1] & 0xff) << 8)
                    + (bufferFrom[curByte] & 0xff)) / (float) Short.MAX_VALUE)
                    * m_hilbert[j];
            }

            // Add the delayed original signal to the Hilbert transform to
            // get the amplitude envelope of the signal
            sTo = ((short)(((bufferFromIn[curByteStartIn + 1] & 0xff) << 8)
                + (bufferFromIn[curByteStartIn] & 0xff))) / (float) Short.MAX_VALUE;
            sFrom = (float) Math.sqrt(sTo * sTo + sFrom * sFrom);
        }

        // Increment buffers if needed
        curByteStart += blockAlign;
```

```
if(curByteStart >= dry.length)
{
    curBufferStart++;
    curByteStart = 0;
    if(curBufferStart < m_storeBuffers.size())
        bufferFrom = m_storeBuffers.get(curBufferStart);
}

// Compress or expand
if (curBufferStartIn >= 0)
{
    // Compute the current amplitude (stored in sFrom) in decibels so
    // that we can compare it to the threshold
    float currentValue = -60F;
    if (Math.abs(sFrom) > 0)
        currentValue = (float) (20D* Math.log10(Math.abs(sFrom)));

    // If the current amplitude is below the threshold, release
    if (currentValue <= m_threshold)
    {
        // If there was compression, then slowly increase the gain.  With
        // each sample, increase the gain by the amount it would have taken
        // to increase the gain from maximum compression (ratio 10:1) over
        // the same time interval for the release, until the gain returns to
        // 1.  Of course, the gain may return to 1 faster than the release
        // time, as the compression ratio could be lower or the compressor
        // may still be "attacking".  Using 10:1 compression here is a
        // simplification
        if (m_compressorGain < 1F)
            m_compressorGain += Math.min(1 - m_compressorGain,
                (1F - (1F / 10F)) / (m_release * format.getSampleRate()
                / 1000F));

        // If there was expansion, do the same, but reduce the gain
        else if (m_compressorGain > 1F)
            m_compressorGain -= Math.min(m_compressorGain - 1,
                (10F - 1F) / (m_release * format.getSampleRate() / 1000F));

        // Do nothing, if both the compression numerator and denominator were
        // equal to 1 (there was no compression or expansion)
    }

    // If the current amplitude is over the threshold, attack.  Unlike the
    // release, the additional (or lower) gain per sample depends on the
    // amount of compression we want to achieve at the end of the attack
    else
    {
```

```
            // If the goal is compression, reduce the gain
            if (m_numerator > 1)
            {
                if (m_compressorGain > (1F / m_numerator))
                    m_compressorGain -= Math.min(m_compressorGain -
                        (1F / m_numerator), (1F - (1F / m_numerator)) /
                        (m_attack * format.getSampleRate() / 1000F));
            }

            // if the goal is expansion, increase the gain
            else if (m_denominator > 1)
            {
                if (m_compressorGain < m_denominator)
                    m_compressorGain += Math.min(-m_compressorGain + m_denominator,
                        (m_denominator - 1F) / (m_attack * format.getSampleRate()
                        / 1000F));
            }
        }

        // Apply the gain to the sample, to which it belongs
        sTo = (short)(((bufferFromIn[curByteStartIn + 1] & 0xff) << 8)
            + (bufferFromIn[curByteStartIn] & 0xff));
        sTo *= m_compressorGain;
    }

    // Increment buffers if needed
    curByteStartIn += blockAlign;
    if(curByteStartIn >= dry.length)
    {
        curBufferStartIn++;
        curByteStartIn = 0;
        if(curBufferStartIn < m_storeBuffers.size())
            bufferFromIn = m_storeBuffers.get(curBufferStartIn);
    }

    // Record the output
    sTo = Math.max(Math.min(sTo, Short.MAX_VALUE), Short.MIN_VALUE);
    wet[i] = (byte)((int) sTo & 0xff);
    wet[i + 1] = (byte)((int) sTo >>> 8 & 0xff);
    }
}
```

13.4. Forward-looking compressor

The example drum compression above works, because the compressor attacks allows the initial hit of the snare to remain and the compressor release makes sure to suppress the ring of the snare. The snare is tigher and more "poppy."

This is not always what a compressor should do. In vocal compression, for example, we may prefer to remove all signal peaks. We can attempt to do so with a shorter attack. If the attack is too long, some peaks will persist. If the attack is too short, the compressor will change not only the amplitude envelope of the underlying wave, but also the wave itself, resulting in distortion. The alternative is to use a *forward-looking compressor*.

The compressor above changes the amplitude of the signal based on its current value. A forward-looking compressor changes the amplitude of the signal based on its expected level in, say, a few milliseconds. This is an important difference. If the attack is positive and the compressor does not look ahead, it will take some time to lower sharp increases in amplitude and a short peak will remain. If the compressor looks ahead, the amplitude of the signal can be lowered before the actual peak occurs and the peak will not remain.

Implementing a forward-looking compressor is simple. Once the amplitude envelope is computed, the gain is applied not to the sample, to which it belongs, but to a previous sample. Suppose that we declare the following variable.

float m_forward – This is the amount of time the compressor looks forward in the future, measured in seconds.

We must also keep track of the where from in the past audio data the value of the original signal is taken, in addition to the indices we already keep for computing the Hilbert transform and the amplitude envelope.

Code 82. Forward looking compression index

```
int curBufferForward = m_storeBuffers.size() - 1;
int curByteForward = (int) ((m_h / 2) + m_forward * format.getSampleRate()
    / 1000) * blockAlign;
curByteForward = -curByteForward;
while (curByteForward < 0)
{
    curByteForward += dry.length;
    curBufferForward--;
}
byte bufferForward [] = null;
if(curBufferForward >= 0)
    bufferForward = m_storeBuffers.get(curBufferForward);
```

We should also adjust how we keep track of used and unused past audio buffers.

Code 83. Past buffers in forward looking compressors

```
while (curBufferStart > 0 && curBufferStartIn > 0 && curBufferForward > 0)
{
    m_storeBuffers.remove(0);
    curBufferStart--;
    curBufferStartIn--;
```

```
    curBufferForward--;
}
```

Finally, we use the forward buffer and byte indices when calculating gain.

Code 84. Applying gain in a forward looking compressor

```
sTo = (short)(((bufferForward[curByteForward + 1] & 0xff) << 8)
    + (bufferForward[curByteForward] & 0xff));
sTo *= m_compressorGain;
```

Here, the current gain is applied to a past sample. This is easier than applying a future gain to the current sample, although it introduces an additional delay in the signal.

13.5. Average vs. peak compression

A compressor that has fast attack and release can compress the peaks of the underlying wave form, acting similarly to "no clip" distortion. We can correct for that, by using an average of the amplitudes calculated with the Hilbert transform over a short period of time. This is not particularly difficult and simply requires the keeping of additional indices for the moving average and, depending on implementation, may introduce an additional delay in the signal.

If, for example, we decide that the we should take the average of the amplitude envelope computed with the Hilbert transform over 0.1 seconds, we simply need to keep track of the rolling average of 0.1 seconds. If we can keep the sum of the required number of samples – 4410 samples in a mono signal at 44100 Hz – then, with each new sample, we can add the value of the Hilbert transform amplitude at that sample and subtract the amplitude of the sample 4410 samples ago. The values of all 4410 samples can be kept in a circular buffer.

13.6. Side chained compressor

A *side chained compressor* changes the amplitude of one signal based on the amplitude of another signal (e.g., the amplitude of the bass based on the amplitude of the drums). The implementation of a side chained compressor is identical to the compressor above, except that the Hilbert transform is applied not to the incoming dry buffer, but to a "control" buffer. The control buffer is named **controlbuffer** in the **apply** function of **EffectInterface** and named simply **control** in the **apply** function in the compressor above.

What the compressor is chained to (i.e., what the control signal is) should be selected by the user through the graphical user interface. In the Orinj effect framework, the side chained compressor should return **true** from the function **allowsSideChaining** of **EffectInterface**. This prompts the Orinj effect dialog to display a drop-down box, where the user can choose the track that provides the signal for the control buffer.

13.7. Multiband compressor

A *multiband compressor* first splits the incoming signal into several frequency bands (e.g., low frequency band between 0 Hz and 300 Hz, mid frequency band between 300 Hz and 3000 Hz, and high frequency band above 3000 Hz). The resulting signals then use independent compressors. The outputs of the compressors are combined into a single output signal.

A multiband compressor can be used as a de-esser. In the three-band compressor described in the previous paragraph, the low and mid frequency bands use no compression or expansion, but a fast attack, large compression ratio compressor is applied to high frequencies. The sound "s", if pronounced in the original signal, is pronounced in the high frequency band. It will be picked up and removed by the compressor. There will be no impacts to low and mid frequencies.

To implement a three-band compressor, we apply three filters to the incoming signal: a low pass filter, a band pass filter, and a high pass filter. We then apply three independent compressors to the three resulting signals and combine the three outputs of the compressors into one. We should be careful to use low pass, band pass, and high pass filters with the same length, so any delays in the three signals are the same and the three signals can be combined. Similarly, we should use Hilbert transforms of similar length, if we are delaying the original signal in any way in order to combine it with the output of the Hilbert transform.

13.8. Multi-threshold compressor

A *multi-threshold compressor*, as the name suggests, uses multiple thresholds and compression ratios. For example, a compressor can leave the signal unchanged if the input is below -20 dB (i.e., the compression ratio is 1:1), compress the signal with a compression ratio of 3:1 if the input is between -20 dB and -10 dB, and compress the signal with a compression ratio of 4:1 if the input is over -10 dB.

A multi-threshold compressor can be implemented with the same code as the one presented above. However, it is not always clear what the compressor should do when compression ratios change. For example, if the input is above -10 dB, but then moves to between -20 dB and -10 dB in our example compressor of the previous paragraph, then the compression ratio should change from 4:1 to 3:1. If the compressor has an attack and a release, does this change represent an attack to a new compression ratio or a release from a higher compression ratio to a lower compression ratio? Should this change take the amount of time represented by the attack or by the release?

Using either of the attack time and the release time works in practice. The designer can make a choice. In Orinj, the compressor always uses the attack time, unless the change is to a compression ratio of 1:1. If the ratio is moving towards 1:1, then the Orinj compressor uses the release time.

13.9. Soft-knee compression

Soft-knee compression makes the transition from one compression ratio to another smoother. In the graph below, there is no compression up to -10 dB, but when the input level exceeds -10 dB, the compression ratio becomes 4:1 (the slope of the compression graph is 1 below -10 dB and ¼ above -10 dB). With hard-knee compression, the switch from 1:1 compression (no compression or expansion) to 4:1 compression is sudden at -10 dB. This is represented by the dotted line in the graph below. With soft-knee compression, the transition is gradual and occurs somewhere between -13 dB and -7 dB. This is represented by the solid line in the graph below.

Figure 12. Soft-knee compression

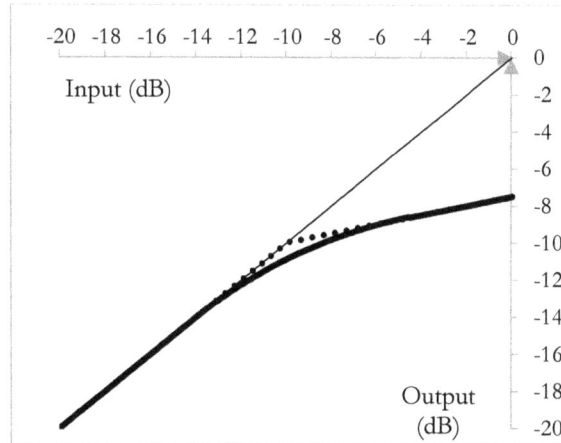

Hard-knee compression changes the compression ratio suddenly, as in the dotted line above.
Soft-knee compression changes the compression ratio gradually, as in the solid line above.

There is nothing different about implementing soft-knee compression. The only difficulty is with figuring out the appropriate compression ratio in the transition interval. For example, what should the compression ratio be at -9 dB?

This is not a DSP problem, but a task in geometry and so we treat it as such below. The following is the code that can be used to draw, in Java, perhaps for the compressor graphical user interface, the transition portion of the graph above. Similar code, with minor adjustments, can be used to get the actual compression ratio in the transition interval. Note though that the actual compression transition could also be something much simpler, such as a straight line.

Suppose that we have two compression segments defined by the input and output points (**input0**, **output0**), (**input1**, **output1**), and (**input2**, **output2**). In the graph above, for example, the two segments are defined by the points (-20, -20), (-10, -10), and (0, -7.5).

Code 85. Soft-knee compression

```
// Recompute the three points of the two segments in terms of coordinates on
// the drawing canvas, rather than dB values
inputvu0 = rec.width * (input0 - Compressor.MINTHRESHOLD) /
    Compressor.MINTHRESHOLD;
inputvu1 = rec.width * (input1 - Compressor.MINTHRESHOLD) /
    Compressor.MINTHRESHOLD;
inputvu2 = rec.width * (input2 - Compressor.MINTHRESHOLD) /
    Compressor.MINTHRESHOLD;

outputvu0 = rec.height + rec.height * (output0 - Compressor.MINTHRESHOLD) /
```

```
   Compressor.MINTHRESHOLD;
outputvu1 = rec.height + rec.height * (output1 - Compressor.MINTHRESHOLD) /
   Compressor.MINTHRESHOLD;
outputvu2 = rec.height + rec.height * (output2 - Compressor.MINTHRESHOLD) /
   Compressor.MINTHRESHOLD;

// The two segment lines are determined by the three points above.  If, for
// example, the first line is output = a0 * input + b0, then a0 and b0 can be
// computed from the equations outputvu0 = a0 * inputvu0 + b0 and
// outputvu1 = a0 * inputvu1 + b0.  We might need some additional work here
// to ensure that the two points (input0, output0) and (input1, output1) are
// not the same, but we ignore this special case here for simplicity

// For the first segment
float a0 = (outputvu1 - outputvu0) / (inputvu1 - inputvu0);
float b0 = outputvu0 - a0 * inputvu0;

// For the second segment
float a1 = (outputvu2 - outputvu1) / (inputvu2 - inputvu1);
float b1 = outputvu1 - a1 * inputvu1;

// Compute the length of the two segments and find out which segment is shorter
float len0 = (float) Math.sqrt((outputvu1 - outputvu0) * (outputvu1 - outputvu0)
   + (inputvu1 - inputvu0) * (inputvu1 - inputvu0));
float len1 = (float) Math.sqrt((outputvu2 - outputvu1) * (outputvu2 - outputvu1)
   + (inputvu2 - inputvu1) * (inputvu2 - inputvu1));
float len = Math.min(len0, len1);

// Suppose that the compressor has a variable knee, which goes from 0 (hard knee)
// to 1 (very soft knee).  Assume that the "softest" knee is a gradual
// transition from the middle of the shorter segment to a point on the longer
// segment that is as far from where the two segments cross as the middle of the
// shorter segment.  Essentially, we want to draw an arc of a circle so that the
// two segments are tangents to the circle.  A smaller knee implies a smaller
// circle.  The following are the tangent points
float midpoint0x = inputvu0 + (inputvu1 - inputvu0)
   * (1F - 0.5F * knee * len / len0);
float midpoint0y = a0 * midpoint0x + b0;
float midpoint1x = inputvu1 + (inputvu2 - inputvu1) * 0.5F * knee * len / len1;
float midpoint1y = a1 * midpoint1x + b1;

// We can now make two lines, each perpendicular to one of the segments and
// crossing the segments at the tangent points.  These two lines cross in the
// center of the circle, the arc of which we want to draw.  If we wanted to draw
// these two lines, we could use the following code.  g is a Java Graphics object
//    float point0x = 0;
//    float point0y = (-1 / a0) * point0x + midpoint0y + (1 / a0) * midpoint0x;
//    float point1x = rec.width;
```

162

```
//      float point1y = (-1 / a0) * point1x + midpoint0y + (1 / a0) * midpoint0x;
//      g.drawLine((int) point0x, (int) point0y, (int) point1x, (int) point1y);
//      point0x = 0;
//      point0y = (-1 / a1) * point0x + midpoint1y + (1 / a1) * midpoint1x;
//      point1x = rec.width;
//      point1y = (-1 / a1) * point1x + midpoint1y + (1 / a1) * midpoint1x;
//      g.drawLine((int) point0x, (int) point0y, (int) point1x, (int) point1y);

// The center of the circle is the point (circlex, circley), computed as the
// intersection of the two lines.  Note that if a0 = a1 (the compression ratios
// of the two segments are the same, we cannot draw a circle and we should just
// draw a straight line.  It may also be useful to round a0 and a1 to some number
// of decimals (e.g., 3) before comparing them
if (a0 != a1)
{
   float circlex = (midpoint0y + (1 / a0) * midpoint0x - midpoint1y - (1 / a1)
      * midpoint1x) / ((1 / a0) - (1 / a1));
   float circley = (-1 / a0) * circlex + midpoint0y + (1 / a0) * midpoint0x;

   // The circle radius is the distance between the intersection and one of the
   // tangent points
   float r = (float) Math.sqrt((circlex - midpoint0x) * (circlex - midpoint0x)
      + (circley - midpoint0y) * (circley - midpoint0y));

   // Draw the circle
   int xfrom = (int) Math.min(midpoint0x, midpoint1x);
   int xto = (int) Math.max(midpoint0x, midpoint1x);
   for(int j = xfrom; j <= xto; j++)
   {
      float yy = 0;
      if (circley > midpoint0y)
         yy = circley - (float) Math.sqrt(r * r - ((float) j - circlex)
            * ((float) j - circlex));
      else
         yy = circley + (float) Math.sqrt(r * r - ((float) j - circlex)
            * ((float) j - circlex));
      g.drawLine(j, (int) yy, j, (int) yy - 1);
   }
}
else
{
   int xfrom = (int) Math.min(midpoint0x, midpoint1x);
   int xto = (int) Math.max(midpoint0x, midpoint1x);
   for(int j = xfrom; j <= inputvu1; j++)
   {
      float yy = a0 * j + b0;
      g.drawLine(j, (int) yy, j, (int) yy - 1);
   }
```

```
    for(int j = (int) inputvu1; j <= xto; j++)
    {
        float yy = a1 * j + b1;
        g.drawLine(j, (int) yy, j, (int) yy - 1);
    }
}
```

Chapter 14. Reverb

Reverbs are discussed in chapter 15 of volume 1. The reverb below is implemented as a sequence of an equalizer, a feedforward comb filter, and three all pass filters, in this order. The equalizer controls the brightness of the reverb by increasing the magnitude of high frequencies and decreasing the magnitude of low frequencies in the reverberated portion of the signal. The comb filter controls a noticeable delay between the original signal and the reverb to create the perception of larger or smaller rooms. The three all pass filters create multiple repetitions of the signal to simulate the late reverb.

14.1. Implementation of the reverb

The code snippet below is long, but uses the same techniques developed for the previous effects. For each of the filters, we store incoming data, keep track of indices, calculate filter output, and remove unused buffers. The implementation of the reverb, while long, is in fact simpler than the implementation of the chorus.

The all pass filters are the first infinite impulse response filters used in this book. These are Shroeder all pass filters that calculate their output based not only on their input data, but also on their past output data. This implies a slight change in how we keep track of indices. Nonetheless, the implementation of these all pass filters is no different than the implementation of any other filter in this book.

The delays of the all pass filters are set to 3 ms, 7 ms, and a larger delay that varies between 8 ms and 25 ms depending on user input as explained below. To make sure that the repetitions introduced by the three all pass filters do not overlap, the delays of the all pass filters are different and should be set to numbers of samples that are mutually prime. For simplicity, they are set to prime numbers, which ensures that they are mutually prime.

The parameters of the reverb implemented below are pre-reverb delay, early decay, length, smoothness, and brightness. The pre-reverb delay is the delay introduced by the comb filter. The early decay is the decay of the comb filter. The length is the length of the reverb created by the all pass filters and is used to calculate the decay of the all pass filters. The smoothness determines the delay of the last all pass filter. This delay varies between 8 ms and 25 ms for a smoother or a less smooth reverb. The brightness of the reverb is determined by a simple two band equalizer that increases the magnitude of high frequencies and decreases the magnitude of low frequencies in the reverberated portion of the signal or vice versa.

The following are the member data of the reverb.

`int [] PRIMES` – This is an array of prime numbers. It includes the numbers from 2, 3, 5, 7, ... to the numbers ..., 3061, 3067, 3079. 3079 samples in 44100 Hz recording is over 0.05 seconds and the array contains more than sufficiently large prime numbers to allow all pass filter delays of up to 25 ms.

float m_predelay – This is the amount of predelay in seconds (the delay of the comb filter). It is between 0.001 seconds and 0.5 seconds.

float m_earlyDecay – This is the amount of early decay (the decay of the comb filter). It is a fraction between 0 and 1.

float m_length – This is the length of the reverb in seconds. It determines the decay of the all pass filters. It is between 0 seconds and 1 second.

float m_smoothness – This is the smoothness of the reverb expressed as a fraction between -1 and 1 (-100% and 100% for the user). It is converted to a delay in the last all pass filter between 25 ms and 8 ms respectively).

float m_brightness – This is the brightness of the reverb. It is expressed as a fraction between 0 and 1 (0% and 100% for the user). It is converted to a gain of -4 dB to 4 dB for high frequencies (4 dB to -4 dB for low frequencies respectively).

float [] m_hhigh – These are the coefficients of the high pass filter used by the equalizer.

float [] m_hlow – These are the coefficients of the low pass filter used by the equalizer.

int FILTERLENGTH – This is the length of the equalizer (the lengths of the high pass and low pass filters).

boolean m_filterSet – This is a flag to note whether the equalizer filters have been computed or should be computed. In principle, since the sampling rate is fixed at 44100 Hz and the cutoff frequency for the equalizer filters is fixed at 4000 Hz for both filters, this flag is unnecessary. The filters can be computed in the constructor of the effect or at the beginning of playback. However, we leave the computation of the filter in the **apply** function, where we have access to the audio format, in case this effect is to be used with different sampling rates.

Vector<byte []> m_storeBuffers – This vector stores incoming audio data for use in the equalizer.

Vector<float []> m_storeEqBuffers – This vector stores the output of the equalizer for use in the comb filter.

Vector<float []> m_storeCombBuffers – This vector stores the output of the comb filter for use in the first all pass filter.

Vector<float []> m_storeAllPassBuffers0 – This vector stores the output of the first all pass filter for use in the second all pass filter and the first all pass filter itself.

Vector<float []> m_storeAllPassBuffers1 – This vector stores the output of the second all pass filter for use in the third all pass filter and the second all pass filter itself.

Vector<float []> m_storeAllPassBuffers2 – This vector stores the output of the third all pass filter for use in this same third all pass filter.

The reverb constructor is as follows.

Code 86. Reverb constructor

```
public Reverb()
{
   // These are the five parameters that the user can change
   m_predelay = (float) 0.05;
   m_earlyDecay = (float) 0.75;
   m_length = (float) 1.0;
   m_smoothness = (float) 0.0;
   m_brightness = (float) 0.5;

   // These are the coefficients of the high pass and low pass filters
   m_hhigh = null;
   m_hlow = null;
   m_filterSet = false;

   // The following are storage at each stage of the reverb computation
   m_storeWetBuffers = new Vector<float []>(0);
   m_storeBuffers = new Vector<byte []>(0);
   m_storeEqBuffers = new Vector<float []>(0);
   m_storeCombBuffers = new Vector<float []>(0);
   m_storeAllPassBuffers0 = new Vector<float []>(0);
   m_storeAllPassBuffers1 = new Vector<float []>(0);
   m_storeAllPassBuffers2 = new Vector<float []>(0);
}
```

At the beginning of playback, the reverb simply initializes all storage.

Code 87. Reverb at the beginning of playback

```
public void startPlay()
{
   m_storeBuffers.setSize(0);
   m_storeWetBuffers.setSize(0);
   m_storeEqBuffers.setSize(0);
   m_storeCombBuffers.setSize(0);
   m_storeAllPassBuffers0.setSize(0);
   m_storeAllPassBuffers1.setSize(0);
   m_storeAllPassBuffers2.setSize(0);
}
```

The implementation of the reverb is as follows. The function **findPrime** finds the prime number in **PRIMES** that is closest to its argument.

Code 88. Reverb

```
public void apply(byte [] dry, byte [] wet, byte [] control, AudioFormat format,
   double time)
```

```
{
    // Compute the filters of the equalizer
    if(! m_filterSet)
        computeFilter(format);

    // Initialize variables
    int blockAlign = (format.getChannels() * format.getSampleSizeInBits()) / 8;
    int channels = format.getChannels();
    float sTo = 0.0F;
    float sFrom = 0.0F;

    // Calculate the gain of the equalizer.  The high pass filter uses gain,
    // whereas the low pass filter uses (1/gain).  The user sets the reverb
    // brightness between 0 and 1.  This computation implies gain between 0.63
    // (-4 dB) and 1.58 (4 dB)
    float gain = (float)Math.pow(10, (m_brightness * 10 - 5) / 25);

    // Calculate the decay of the all pass filters.  0.001 is the gain equivalent
    // to -60 dB or silence.  If the repetition delay is 7 ms, this silence must
    // be achieved in (m_length / 0.007) repetitions and so the decay must be
    // 0.001^(1 / (m_length / 0.007)).  Additional code is needed to ensure that
    // there are no problems if the length is zero.  Technically, the first
    // statement below ignores the fact that there is an all pass filter with a
    // larger delay below (the one that is adjusted for smoothness), but adjusting
    // for that is easy
    float decay = (float) Math.pow(0.001, 0.007 / m_length);

    // Store incoming audio data for use by the equalizer
    byte [] storeBuffer = new byte [dry.length];
    System.arraycopy(dry, 0, storeBuffer, 0, dry.length);
    m_storeBuffers.add(storeBuffer);

    // The output of the equalizer
    float weteq[] = new float[(dry.length * channels) / blockAlign];

    // Compute how far back the equalizer should look, given FILTERLENGTH
    int curBufferStart = m_storeBuffers.size() - 1;
    int curByteStart = FILTERLENGTH * blockAlign;
    curByteStart = -curByteStart;
    while (curByteStart < 0)
    {
        curByteStart += dry.length;
        curBufferStart--;
    }
    byte bufferFromStart[] = null;
    if(curBufferStart >= 0)
        bufferFromStart = m_storeBuffers.get(curBufferStart);
```

```
// Remove unused buffers
while (curBufferStart > 0)
{
   m_storeBuffers.remove(0);
   curBufferStart--;
}

// Compute the output of the equalizer at each sample
for(int i = 0, weti = 0; i < dry.length; i += blockAlign,
   curByteStart += blockAlign, weti += channels)
{
   // Increment buffers if needed
   if(curByteStart >= dry.length)
   {
      curBufferStart++;
      bufferFromStart = m_storeBuffers.get(curBufferStart);
      curByteStart = 0;
   }

   // Initialize variables
   int curBuffer = curBufferStart;
   int curByte = curByteStart;
   byte bufferFrom[] = bufferFromStart;
   sTo = 0.0F;

   // Compute the output of the equalizer at one sample
   for(int j = 0; j < FILTERLENGTH; j++, curByte += blockAlign)
   {
      if(curByte >= dry.length)
      {
         curBuffer++;
         bufferFrom = m_storeBuffers.get(curBuffer);
         curByte = 0;
      }

      if(curBufferStart >= 0)
      {
         sFrom = (short)(((bufferFrom[curByte + 1] & 0xff) << 8)
            + (bufferFrom[curByte] & 0xff));
         sTo += sFrom * m_hhigh[j] * gain;
         sTo += sFrom * m_hlow[j] / gain;
      }
   }

   // Store the output of the equalizer
   sTo = Math.max(Math.min(sTo, Short.MAX_VALUE), Short.MIN_VALUE);
   weteq[weti] = sTo;
```

```
}

// Store the output of the equalizer for use by the comb filter
m_storeEqBuffers.add(weteq);

// We need a buffer for floating points, as from here on we work
// with floating-point audio data, not byte arrays
float bufferFrom[] = null;

// Since we did not copy the data of wetf, but simply added them to the
// the vector of equalizer output buffers, we need another one.  This
// is the output of the comb filter
float wetComb[] = new float[(dry.length * channels) / blockAlign];

// Compute how far back the comb filter should look
int curBuffer = m_storeEqBuffers.size() - 1;
int curByte = (int) (m_predelay * format.getSampleRate() * channels);
while (curByte > 0 && curByte % channels != 0)
    curByte--;
curByte = Math.max(curByte, 0);
curByte = -curByte;
while (curByte < 0)
{
    curByte += weteq.length;
    curBuffer--;
}

// We can do without the variable below and we can remove unused buffers
// here, but using this variable is cleaner.  This variable is used
// below to remove unused buffers
int firstUsedBuffer = curBuffer;

// Compute the output of the comb filter at each sample
for(int i = 0; i < weteq.length; i += channels, curByte += channels)
{
    sTo = 0;

    if (curBuffer >= 0)
    {
        // Increment buffers as needed.  We use weteq here, as bufferFrom could
        // still be null
        if (curByte >= weteq.length)
        {
            curBuffer++;
            curByte = 0;
        }
        bufferFrom = m_storeEqBuffers.get(curBuffer);
```

```
        // Compute the output of the comb filter at one sample.  Note that
        // we do not use the original signal, but only the repetition, as we
        // simply want to delay the whole reverb
        sTo = (int) ((float) bufferFrom[curByte] * m_earlyDecay);
        wetComb[i] = sTo;
    }
}

// Remove unused buffers
while (firstUsedBuffer > 0)
{
    m_storeEqBuffers.remove(0);
    firstUsedBuffer--;
}

// Store the output of the comb filter for use in the first all pass filter
m_storeCombBuffers.add(wetComb);

// This is the output of the first all pass filter.  Store it for use in the
// second all pass filter (and in the first all pass filter itself)
float [] wetAllPass0 = new float [(dry.length * channels) / blockAlign];
m_storeAllPassBuffers0.add(wetAllPass0);

// Since the all pass filter is an IIR filter, it uses not only past
// incoming data, but also past data from its own output
float [] bufferFromAllPass = null;

// Compute how far back the first all pass filter should look.  Note the use
// of a prime number of samples approximately equal to 7 ms of audio data
curBuffer = m_storeCombBuffers.size() - 1;
curByte = findPrime((int) (0.007F * format.getSampleRate())) * channels;
while (curByte > 0 && curByte % channels != 0)
    curByte--;
curByte = Math.max(curByte, 0);
curByte = -curByte;
while (curByte < 0)
{
    curByte += wetComb.length;
    curBuffer--;
}
if (curBuffer >= 0)
{
    bufferFrom = m_storeCombBuffers.get(curBuffer);
    bufferFromAllPass = m_storeAllPassBuffers0.get(curBuffer);
}

// Keep track of which buffers are used
```

```java
int firstUsedBufferComb = curBuffer;

// Compute the output of the first all pass filter at each sample
for(int i = 0; i < weteq.length; i += channels, curByte += channels)
{
    if (curBuffer >= 0)
    {
        // Increment buffers if needed
        if (curByte >= weteq.length)
        {
            curBuffer++;
            curByte = 0;
            bufferFrom = m_storeCombBuffers.get(curBuffer);
            bufferFromAllPass = m_storeAllPassBuffers0.get(curBuffer);
        }

        // Compute the output of the first all pass filter at one sample
        sTo = (int) (decay * wetComb[i] + bufferFrom[curByte]
            - decay * bufferFromAllPass[curByte]);
        wetAllPass0[i] = sTo;
    }
    else
    {
        // At the beginning of playback, where we do not have enough data
        // to compute the all pass filter, set the output of the all pass
        // filter to the comb filter
        wetAllPass0[i] = wetComb[i];
    }
}

// Keep track of which buffers are used
int firstUsedBufferAllPass0 = firstUsedBufferComb;

// Remove unused buffers
while (firstUsedBufferComb > 0)
{
    m_storeCombBuffers.remove(0);
    firstUsedBufferComb--;
}

// This is the output of the second all pass filter.  We declare a new buffer
// as the buffer used for the first all pass filter is needed and we should
// not overwrite it
float wetAllPass1[] = new float[(dry.length * channels) / blockAlign];
m_storeAllPassBuffers1.add(wetAllPass1);
bufferFromAllPass = null;

// Compute how far back the second all pass filter should look.  Use a prime
```

```java
// number of samples approximately equal to 3 ms of audio data
curBuffer = m_storeAllPassBuffers0.size() - 1;
curByte = findPrime((int) (0.003F * format.getSampleRate())) * channels;
while (curByte > 0 && curByte % channels != 0)
    curByte--;
curByte = Math.max(curByte, 0);
curByte = -curByte;
while (curByte < 0)
{
    curByte += wetAllPass0.length;
    curBuffer--;
}
if (curBuffer >= 0)
{
    bufferFrom = m_storeAllPassBuffers0.get(curBuffer);
    bufferFromAllPass = m_storeAllPassBuffers1.get(curBuffer);
}

// Keep track of which buffers are used
firstUsedBufferAllPass0 = Math.min(curBuffer, firstUsedBufferAllPass0);
int firstUsedBufferAllPass1 = curBuffer;

// Compute the output of the second all pass filter
for(int i = 0; i < wetf.length; i += channels, curByte += channels)
{
    if (curBuffer >= 0)
    {
        // Increment buffers if needed
        if (curByte >= wetf.length)
        {
            curBuffer++;
            curByte = 0;
            bufferFrom = m_storeAllPassBuffers0.get(curBuffer);
            bufferFromAllPass = m_storeAllPassBuffers1.get(curBuffer);
        }

        // Compute the output of the second all pass filter at one sample
        sTo = (int) (decay * wetAllPass0[i] + bufferFrom[curByte]
            - decay * bufferFromAllPass[curByte]);
        wetAllPass1[i] = sTo;
    }
    else
    {
        sTo = (int) wetAllPass0[i];
        wetAllPass1[i] = sTo;
    }
}
```

```java
// This is the output of the third all pass filter
float wetAllPass2[] = new float[(dry.length * channels) / blockAlign];
m_storeAllPassBuffers2.add(wetAllPass2);
bufferFromAllPass = null;

// Compute how far back the third all pass filter should look to
// obtain data, but vary that delay based on the smoothness of the reverb.
// The reverb smoothness varies between -1 and 1, which varies the delay
// between 25 milliseconds and 8 milliseconds respectively
curBuffer = m_storeAllPassBuffers1.size() - 1;
curByte = findPrime((int) (((0.008 + 0.025) / 2 - ((0.025 - 0.008)
    * m_smoothness / 2)) * format.getSampleRate())) * channels;
while (curByte > 0 && curByte % channels != 0)
    curByte--;
curByte = Math.max(curByte, 0);
curByte = -curByte;
while (curByte < 0)
{
    curByte += wetAllPass1.length;
    curBuffer--;
}
if (curBuffer >= 0)
{
    bufferFrom = m_storeAllPassBuffers1.get(curBuffer);
    bufferFromAllPass = m_storeAllPassBuffers2.get(curBuffer);
}

// Keep track of which buffers are used
firstUsedBufferAllPass1 = Math.min(curBuffer, firstUsedBufferAllPass1);
int firstUsedBufferAllPass2 = curBuffer;

// Compute the output of the third all pass filter
for(int i = 0; i < dry.length; i += blockAlign, curByte += channels)
{
    if (curBuffer >= 0)
    {
        // Increment buffers if needed
        if (curByte >= wetf.length)
        {
            curBuffer++;
            curByte = 0;
            bufferFrom = m_storeAllPassBuffers1.get(curBuffer);
            bufferFromAllPass = m_storeAllPassBuffers2.get(curBuffer);
        }

        // The output of the third all pass filter is the output of the
        // effect
        sTo = (int) (decay * wetAllPass1[i * channels / blockAlign]
```

```
                   + bufferFrom[curByte] - decay * bufferFromAllPass[curByte]);
           wetAllPass2[i * channels / blockAlign] = sTo;
           sTo = Math.min(Math.max(sTo, Short.MIN_VALUE), Short.MAX_VALUE);
           wet[i] = (byte) ((int) sTo & 0xff);
           wet[i + 1] = (byte) ((int) sTo >>> 8 & 0xff);
        }
        else
        {
           sTo = (int) wetAllPass1[i * channels / blockAlign];
           wetAllPass2[i * channels / blockAlign] = sTo;
           sTo = Math.min(Math.max(sTo, Short.MIN_VALUE), Short.MAX_VALUE);
           wet[i] = (byte) ((int) sTo & 0xff);
           wet[i + 1] = (byte) ((int) sTo >>> 8 & 0xff);
        }
     }

     // Remove unused buffers
     while (firstUsedBufferAllPass0 > 0)
     {
        m_storeAllPassBuffers0.remove(0);
        firstUsedBufferAllPass0--;
     }

     while (firstUsedBufferAllPass1 > 0)
     {
        m_storeAllPassBuffers1.remove(0);
        firstUsedBufferAllPass1--;
     }

     while (firstUsedBufferAllPass2 > 0)
     {
        m_storeAllPassBuffers2.remove(0);
        firstUsedBufferAllPass2--;
     }
  }
```

14.2. Properties of the digital reverb

Some reverb is needed. Without it, a sound recording may sound dry and boring. Reverb makes music livelier as this is what it is meant to do – to simulate the real-life effects of a sound in a room, a club, or a concert hall. Equally important, most music that we listen to already has artificial reverberations and that is what listeners are used to. Music would sound "unnatural" without the artificial reverb.

Older recordings simulated reverb in various mechanical ways – using chamber, spring, or plate reverb units. Some of these are still used today.

A *chamber reverb unit* attempts to record the natural reverb in a room by picking up the sound in the room with microphones placed around to pick up that natural reverb of the room. The chamber reverb unit is a simple concept, but can have many variations – various rooms, room treatments, types and number of microphones, microphone placements, or sound origin placements. While the chamber reverb concept is simple, applying it in practice is not. A big problem is obviously transport. Once everything is set up, it is difficult to move the specific setup to a different place.

A *plate reverb unit* is probably the oldest "unnatural" way to simulate a "natural" reverb. A plate reverb unit directs the sound towards a metal plate. The metal plate vibrates with the sound. The metal plate vibrations, which include the original sound vibrations and some residual ones (the reverb) are then picked up. To accomplish this task, a plate reverb unit uses *transducers*. A transducer is simply a piece of equipment that transforms one type of energy into another. A speaker driver is a transducer, since it transforms electrical energy into mechanical energy as running a signal through the speaker driver makes it move back and forth. A guitar pickup is another type of transducer. It transforms the mechanical energy of the guitar string into an electrical signal. Microphones are also transducers. A plate reverb unit can use a transducer similar to a speaker driver to induce vibrations in the metal plate from the electrical signal that represents the sound. Another transducer – one that works as a pickup – can pick up the vibrations of the plate. The second transducer will pick up not only the original plate vibrations but also the residual plate vibrations creating in this way a "reverberated" sound.

Plate reverbs were used early (say 1950s), but they are still very relevant now. Many digital reverbs actually simulate various plate reverbs. A lot of digital simulators of plate reverbs are smooth (i.e., they have large diffusion) and have a specific coloration, accentuating frequencies over 2 KHz. This means that this reverb accentuates frequencies characteristic for metal clanks, for sibilance, and for shimmer.

Spring reverb units are very common. A lot of guitar amplifiers have strings that vibrate to create reverb. The spring reverb works similarly to the plate reverb. A spring reverb unit uses a transducer to move a spring (or springs) and uses another transducer to pick up the spring vibrations. The residual spring vibrations simulate reverb. Spring reverbs tend to be less smooth than plate reverbs and to have a different coloration. They accentuate the mid-range frequencies and do not create similar sibilance and shimmer. They also have more pronounced early reflections.

The power of digital reverbs comes from the ability of the user to control the reverb characteristics – brightness, total length, early reflection decay, and so on. The typical controls of the artificial digital reverb are similar to the ones of the natural reverb, although they may be named differently. Rather than *brightness* we might see *coloration*, *equalization*, or *high pass cutoff*. Rather than *length* we may see *room dimension*. Rather than *smoothness*, we may see *irregularity* or *left / right perception*.

Artificial reverbs may also have "unnatural" controls. For example, we can have a very pronounced reverb, but one which cuts off quickly rather than decaying smoothly. This can be implemented by running the reverberated portion of the sound through a noise gate. A reverb such as this, with a *gated decay threshold*, can be implemented either with an actual noise gate, or with an expander. An expander, rather than completely disallowing the signal, will attenuate the signal when it falls below the threshold.

A reverb with *gated time*, rather than disallowing the reverb based on an amplitude threshold, would cut it off based on the amount of decay time. For example, we may want a reverb that stops 250 ms after it starts, independently of whether the reverberations quieted down or not. This type of digital reverb is rare, but can be implemented by first compressing the original sound to make its dynamics more even, and then applying a reverb with a gated decay threshold.

A *preverb* is a reverb that, rather than decaying after the original signal, builds up to the original signal. Sometimes a preverb is called a "reverse reverb." What is interesting about this type of reverb is that it is technically *non-causal* in the time domain. A signal operation is *causal in the time domain*, if its output depends only on the previous or current values of the input and not on future values (an operation or a signal is *causal in the frequency domain*, if it does not use negative frequencies). To make this operation causal and to be able to easily implement the preverb, one can create a copy of the input signal and shift that copy back in time.

Finally, reverb is a spatial phenomenon. It is useful to be able to accentuate or modify how the reverb works in the physical space. One way to do that is to bounce the reverb between the two channels of a stereo recording – left and right, although this may sound too unnatural after some time. It may be better to differentiate between the reverbs applied to the left and right channels. We can, for example use reverb that has larger length but starts with lower amplitude in one of the channels. A reverb with 1250 ms total length on the right and 1100 ms total length on the left with minor changes in the initial amplitude is an interesting effect. It provides a better spatial position to the listener, rather than the sound origin, the position of which is defined with panning. In this case, however, to avoid an effect similar to a slap back delay or a flanger, we must ensure that the digital reverb implementation does not mix the signal in the left and right channels.

Chapter 15. Wah wah

A *wah wah* is a sound effect that imitates the sound of the syllable "wah." The wah wah below deploys a shelving filter, but can also be implemented with a band pass filter. The shelving filter used here is discussed in chapter 19 of volume 1. It passes a narrow band of frequencies. The wah wah sound is created when this band begins at high frequencies and gradually moves mid frequencies, then low frequencies, and then back mid and high frequencies. The output of the filter is mixed with the original signal.

Moving a band pass filter down and then up the frequency spectrum is difficult. The filter coefficients must be recomputed constantly. A finite impulse response filter will make the wah wah code slow. An infinite impulse response filter is faster, but recomputing the coefficients of such a filter will make the filter unstable and the amplitude of the resulting sound will increase indefinitely. To speed up the filter and make it stable, we use an infinite impulse response filter on segments of the signal, which we will call frames. The resulting frames of output are then mixed into a single signal.

The following figure shows the magnitude response of the wah wah filter. It should help visualize the filter. This filter was created with the cutoff frequency $\omega_c = 1.6$, which, at the sampling rate 44100 Hz, translates to about 11250 Hz. The cutoff frequency obviously changes as the wah wah progresses. The filter width is 0.2, or about 1400 Hz. The filter gain is 1000, or about 60 dB.

Figure 13. Magnitude response of the wah wah filter

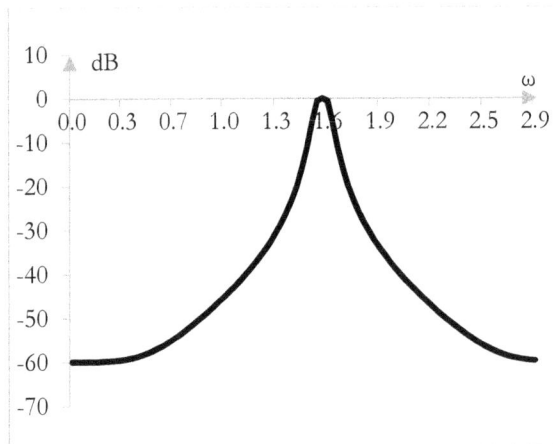

The wah wah filter is a peak filter or a narrow band pass filter. It passes a narrow band of frequencies, but attenuates all other frequencies. In this figure, the narrow band is centered around the normalized angular frequency 1.60, but this center frequency changes as the wah wah time progresses.

The member data of the wah wah are as follows.

int FRAME – This is the size of the frames on which the filters is applied. The frame size is set to 1024.

float m_attack – This is the time that it takes the wah wah band pass filter to move from high frequencies down to low frequencies. It is measured in seconds and can be between 0 seconds and 1 second.

float m_sustainDown – This is the time that the wah wah band pass filter spends at the low frequencies. It is measured in seconds and can be between 0 seconds and 1 second.

float m_release – This is the time it takes the wah wah band pass filter to move from low frequencies back to high frequencies. It is measured in seconds and can be between 0 seconds and 1 second.

float m_sustainUp – This is the time the wah wah band pass filter spends at the high frequencies. It is measured in seconds and can be between 0 seconds and 1 second.

float m_top – This is the highest frequency of the midpoint of the wah wah band pass filter. It is measured in Hz and can be between 1000 Hz and 11000 Hz.

float m_bottom – This is the lowest frequency of the midpoint of the wah wah band pass filter. It is measured in Hz and can be between 0 and 1000 Hz.

float m_filterWidth – This is the width of the band pass filter. It is measured in Hz and can be between 0 and 1000 Hz.

double [] m_storeDry0 – The wah wah uses an infinite impulse response filter of five coefficients. The filter uses the past five samples of the original signal. Rather than storing the whole incoming audio buffers, the wah wah stores only the past five samples of data.

double [] m_storeWet0 – Since the wah wah filter is an infinite impulse response filter, it must also use the past samples of its own output. These samples are stored here.

double [] m_storeDry1 – The wah wah filter is applied to consecutive frames of audio data, but these frames must overlap so that they can be crossfaded and mixed. At any point, the wah wah applies two band pass filters – one on the previous frame and one on the current frame. This buffer stores the past samples of the original signal for the second filter.

double [] m_storeWet1 – This stores the output of the second band pass filter for use by that same filter.

double [] m_a0 – These are the coefficients of the numerator of the first band pass filter.

double [] m_b0 – These are the coefficients of the denominator of the first band pass filter.

double [] m_a1 – These are the coefficients of the numerator of the second band pass filter.

double [] m_b1 – These are the coefficients of the denominator of the second band pass filter.

long m_frameCount – This is a counter that is used to determine when a frame ends and the next frame begins.

The following is the wah wah constructor.

Code 89. Wah wah constructor

```
public Wah()
{
    // These are the parameters that the user can change
    m_attack = 0.5F;
    m_sustainDown = 0.5F;
    m_release = 0.5F;
    m_sustainUp = 0.5F;
    m_top = 10000;
    m_bottom = 500;
    m_filterWidth = 50;

    // Initialize storage for two consecutive filter segments
    m_storeDry0 = new double [4];
    m_storeWet0 = new double [4];
    m_storeDry1 = new double [4];
    m_storeWet1 = new double [4];

    // Initialize the filter numerator and denominator coefficients for two
    // consecutive filter segments
    m_a0 = new double [5];
    m_b0 = new double [5];
    m_a1 = new double [5];
    m_b1 = new double [5];
    m_a0[0] = m_a0[1] = m_a0[2] = m_a0[3] = m_a0[4] = 0;
    m_b0[0] = m_b0[1] = m_b0[2] = m_b0[3] = m_b0[4] = 0;
    m_a1[0] = m_a1[1] = m_a1[2] = m_a1[3] = m_a1[4] = 0;
    m_b1[0] = m_b1[1] = m_b1[2] = m_b1[3] = m_b1[4] = 0;

    // Initialize the counter for the samples in a segment
    m_frameCount = 0;
}
```

The following code is executed at the beginning of playback.

Code 90. Wah wah at the beginning of playback

```
public void startPlay()
{
    m_storeDry0[0] = m_storeDry0[1] = m_storeDry0[2] = m_storeDry0[3] = 0;
    m_storeWet0[0] = m_storeWet0[1] = m_storeWet0[2] = m_storeWet0[3] = 0;
    m_storeDry1[0] = m_storeDry1[1] = m_storeDry1[2] = m_storeDry1[3] = 0;
    m_storeWet1[0] = m_storeWet1[1] = m_storeWet1[2] = m_storeWet1[3] = 0;
```

```
    m_frameCount = 0;
}
```

The following is the implementation of the wah wah.

Code 91. Wah wah

```
public void apply(byte [] dry, byte [] wet, byte [] control, AudioFormat format,
    double time)
{
    // If the whole period of the wah wah is zero, there is no wah wah
    if (m_attack == 0 && m_sustainDown == 0 && m_release == 0 && m_sustainUp == 0)
    {
        for(int i = 0; i < dry.length; i++)
            wet[i] = 0;
        return;
    }

    // Initialize variables
    int blockAlign = (format.getChannels() * format.getSampleSizeInBits()) / 8;
    double sFrom = 0.0F;
    double sTo = 0.0F, sTo0 = 0, sTo1 = 0;

    // This variable is used to compute where in the wah wah period we are -
    // during the attack, release, or sustains
    double curtime = 0;

    // This is the current midpoint of the band pass filter.  During the effect,
    // this midpoint begins at a high frequency, shifts down to lower
    // frequencies and then back up to higher frequencies
    double frequency = 0;

    // This is the midpoint of the band pass filter, but expressed as an
    // angular frequency and used in the computations of the filter coefficients
    double wc = 0;

    // This is the width of the band pass filter
    double B = 0;

    // This is the gain of the shelving filter.  In this implementation, the gain
    // is set to -60 dB.  This effectively makes the shelving filter a band pass
    // filter.  It is possible to use lower gain in absolute value, but the wah
    // wah will be less audible then.  The filter is first created as a shelving
    // filter that boosts a band of frequencies by gain.  The filter numerator
    // coefficients are then scaled down by (1 / gain) to convert the filter into
    // a band pass one
    double gain = 1F / (float)Math.pow(10D, -60F / 20F);

    // Compute the wah wah
```

```java
for(int i = 0; i < dry.length; i += blockAlign, m_frameCount++)
{
    // Move to the next frame.  Frames overlap by half of their length.
    // This means that there are two filters at each sample - one for
    // the previous frame and one for the current frame
    if (m_frameCount % (FRAME / 2) == 0)
    {
        // Calculate the current time.  We need to know whether we are during
        // the attack, release, or sustains of the wah wah
        curtime = time + ((double) i / blockAlign) / format.getSampleRate();
        curtime -= ((int) (curtime / (m_attack + m_sustainDown + m_release
            + m_sustainUp))) * (m_attack + m_sustainDown + m_release
            + m_sustainUp);

        // Depending on where we are in the wah wah cycle, determine the
        // midpoint of the band pass filter
        if (curtime < m_attack)
            frequency = m_top - (m_top - m_bottom) * curtime / m_attack;
        else if (curtime < m_attack + m_sustainDown)
            frequency = m_bottom;
        else if (curtime < m_attack + m_sustainDown + m_release)
            frequency = m_bottom + (m_top - m_bottom) * (curtime - m_attack
                - m_sustainDown) / (m_release);
        else
            frequency = m_top;

        // If we are not in the first frame, switch the two filters.  The filter
        // with index 0 applies to the previous frame.  The filter with index
        // 1 applies to the current frame.  Copy the coefficients of the filter
        // with index 1 over those of the filter with index 0 and, below,
        // compute new coefficients for the filter with index 1
        if (m_frameCount > 0)
        {
            System.arraycopy(m_a1, 0, m_a0, 0, 5);
            System.arraycopy(m_storeDry1, 0, m_storeDry0, 0, 4);
        }
        else
        {
            // If this is the first frame, remove the filter with index 0.
            // This can also be done at the start of playback
            m_a0[0] = m_a0[1] = m_a0[2] = m_a0[3] = m_a0[4] = 0;
            m_b0[0] = m_b0[1] = m_b0[2] = m_b0[3] = m_b0[4] = 0;
            m_storeDry0[0] = m_storeDry0[1] = m_storeDry0[2]
                = m_storeDry0[3] = 0;
            m_storeWet0[0] = m_storeWet0[1] = m_storeWet0[2]
                = m_storeWet0[3] = 0;
        }
```

```
// Convert the midpoint of the band pass filter into an angular
// frequency
wc = 2 * Math.PI * frequency / format.getSampleRate();

// Convert the width of the pass band of the filter into angular
// frequencies.  This could be done only once at the beginning
// of playback
B = 2 * Math.PI * m_filterWidth / format.getSampleRate();

// Compute the coefficients of the filter
m_a1[0] = 16 + wc * wc * wc * wc + 8 * wc * wc + 8 * Math.sqrt(2 * gain)
    * B + 2 * Math.sqrt(2 * gain) * B * wc * wc + 4 * gain * B * B;
m_a1[1] = -64 + 4 * wc * wc * wc * wc - 16 * Math.sqrt(2 * gain) * B
    + 4 * Math.sqrt(2 * gain) * B * wc * wc;
m_a1[2] = 96 + 6 * wc * wc * wc * wc - 16 * wc * wc - 8 * gain * B * B;
m_a1[3] = -64 + 4 * wc * wc * wc * wc + 16 * Math.sqrt(2 * gain) * B
    - 4 * Math.sqrt(2 * gain) * B * wc * wc;
m_a1[4] = 16 + wc * wc * wc * wc + 8 * wc * wc - 8 * Math.sqrt(2 * gain)
    * B - 2 * Math.sqrt(2 * gain) * B * wc * wc + 4 * gain * B * B;
m_b1[0] = 16 + wc * wc * wc * wc + 8 * wc * wc + 8 * Math.sqrt(2) * B
    + 2 * Math.sqrt(2) * B * wc * wc + 4 * B * B;
m_b1[1] = -64 + 4 * wc * wc * wc * wc - 16 * Math.sqrt(2) * B
    + 4 * Math.sqrt(2) * B * wc * wc;
m_b1[2] = 96 + 6 * wc * wc * wc * wc - 16 * wc * wc - 8 * B * B;
m_b1[3] = -64 + 4 * wc * wc * wc * wc + 16 * Math.sqrt(2) * B
    - 4 * Math.sqrt(2) * B * wc * wc;
m_b1[4] = 16 + wc * wc * wc * wc + 8 * wc * wc - 8 * Math.sqrt(2) * B
    - 2 * Math.sqrt(2) * B * wc * wc + 4 * B * B;

// Scale all filter coefficients so that the first coefficient in the
// denominator is 1 (m_b1[0] = 1).  This scaling allows us to apply
// the filter (to create its impulse response from the transfer
// function).  m_b1[0] is not used when applying the filter and is not
// scaled below.  In addition, scale the numerator coefficients by the
// gain so that the shelving filter that boosts a band of frequencies
// and leaves the magnitude of other frequencies unchanged becomes a
// band pass filter that passes that same band are reduces the magnitude
// of the other frequencies
m_a1[0] /= m_b1[0] * gain;
m_a1[1] /= m_b1[0] * gain;
m_a1[2] /= m_b1[0] * gain;
m_a1[3] /= m_b1[0] * gain;
m_a1[4] /= m_b1[0] * gain;
m_b1[1] /= m_b1[0];
m_b1[2] /= m_b1[0];
m_b1[3] /= m_b1[0];
m_b1[4] /= m_b1[0];
```

```
    // Initialize storage for the new filter.  It is not safe for
    // infinite impulse response filters to start with nonzero data
    m_storeDry1[0] = m_storeDry1[1] = m_storeDry1[2] = m_storeDry1[3] = 0;
    m_storeWet1[0] = m_storeWet1[1] = m_storeWet1[2] = m_storeWet1[3] = 0;
}

// The output of the two band pass filters will be crossfaded.  These
// are the corresponding gains
double crossfadeGain0 = 0.5;
double crossfadeGain1 = 0.5;

if (m_frameCount < FRAME / 2)
{
    // At the beginning of playback, use only the filter with index 1
    crossfadeGain0 = 0;
    crossfadeGain1 = 1;
}
else
{
    // During playback, after the first frame, set the gains depending on
    // how far into the frame we are
    long current = m_frameCount - ((m_frameCount / (FRAME / 2))
        * (FRAME / 2));
    crossfadeGain0 = 1 - (double) current / (FRAME / 2);
    crossfadeGain1 = (double) current / (FRAME / 2);
}

// Apply both filters to the current sample
sFrom = (short)(((dry[i + 1] & 0xff) << 8) + (dry[i] & 0xff));
sTo0 = m_a0[0] * sFrom + m_a0[1] * m_storeDry0[3]
    + m_a0[2] * m_storeDry0[2] + m_a0[3] * m_storeDry0[1]
    + m_a0[4] * m_storeDry0[0] - m_b0[1] * m_storeWet0[3]
    - m_b0[2] * m_storeWet0[2] - m_b0[3] * m_storeWet0[1]
    - m_b0[4] * m_storeWet0[0];
sTo1 = m_a1[0] * sFrom + m_a1[1] * m_storeDry1[3]
    + m_a1[2] * m_storeDry1[2] + m_a1[3] * m_storeDry1[1]
    + m_a1[4] * m_storeDry1[0] - m_b1[1] * m_storeWet1[3]
    - m_b1[2] * m_storeWet1[2] - m_b1[3] * m_storeWet1[1]
    - m_b1[4] * m_storeWet1[0];

// Move storage by one sample.  Note that sFrom is placed in the storage
// of the dry incoming data, while sTo is placed in the storage of the wet
// output
System.arraycopy(m_storeDry0, 1, m_storeDry0, 0, 3);
m_storeDry0[3] = sFrom;
System.arraycopy(m_storeWet0, 1, m_storeWet0, 0, 3);
```

```java
            m_storeWet0[3] = sTo0;
            System.arraycopy(m_storeDry1, 1, m_storeDry1, 0, 3);
            m_storeDry1[3] = sFrom;
            System.arraycopy(m_storeWet1, 1, m_storeWet1, 0, 3);
            m_storeWet1[3] = sTo1;

            // Crossfade
            sTo = sTo0 * crossfadeGain0 + sTo1 * crossfadeGain1;

            // Place the result in the output
            sTo = Math.max(Math.min(sTo, Short.MAX_VALUE), Short.MIN_VALUE);
            wet[i] = (byte)((int)sTo & 0xff);
            wet[i + 1] = (byte)((int)sTo >>> 8 & 0xff);
        }
    }
```

Chapter 16. Pitch shift

To change the pitch shift of a wave while preserving its tempo, we take the *discrete Fourier transform (DFT)* of the audio data. This allows us to understand the magnitude of frequencies present in the signal. We then compute the magnitudes that should be present in a pitch shifted signal and use the inverse discrete Fourier transform to produce the pitch shifted signal. The forward and inverse Fourier transform are discussed in chapter 12 of volume 1.

Since the frequency content of a signal changes over time, we take the Fourier transform of chunks of the signal. The corresponding chunks of the pitch shifted signal must be combined. We make sure that the chunks overlap and we crossfade the output signal, just as we did with the wah wah in the previous chapter.

16.1. Actual frequencies

The DFT is not precise, because it has a finite number of components. Take a DFT of 40 components over the sampling frequency 2000. The components are 0 Hz, 50 Hz, 100 Hz, …, 950 Hz. The frequency 50 Hz with peak amplitude 1 after the DFT will show up at the second component (50 Hz) with magnitude 1. However, the frequency 75 Hz, which does not fall exactly on a DFT component, will show up after the DFT at component 0 (0 Hz) with magnitude 0.05, component 1 (50 Hz) with magnitude 0.513, component 2 (100 Hz) with magnitude 0.727 and so on.

By monitoring the phases of the frequencies in the DFT components, it is possible to approximate the actual frequencies present in a signal. For example, if we are to take the DFT again 10 samples later, we will find that the phase of the DFT component 1 has changed, but not by as much as expected. If the frequency at the 50 Hz component was truly 50 Hz, then the change in phase in that frequency after 10 samples should be $-2\pi\, 10 * 50 / 2000 = -\pi/2$. If the frequency at the component 50 Hz is 75 Hz, then its phase after 10 samples will change by $-2\pi\, 10 * 75 / 2000 = -3\pi/4$. This difference of $\pi/4$ tells us that the actual frequency is 50 Hz $+ (\pi/4) * 2000 / (2\pi\, 10) = 75$ Hz.

In practice, the DFT is less precise. The differences between the expected change phase at 50 Hz and the actual change in phase will be computed as 0.659, which implies an actual frequency of approximately 71 Hz. Similar computations at the 100 Hz DFT component will produce the frequency 79 Hz.

16.2. Overlap

In the example above, rather than taking the DFT of 40 components at each consecutive 40 samples, we use overlapping segments. We computed changes in phase and actual frequencies after 10 samples, which implies an overlap of 75%. The larger the overlap, the more likely that the computed actual frequency is close to the DFT component frequency.

The overlap also allows us to crossfade the pitch shifted signal, similarly to the crossfading used in the wah wah. This avoids discontinuities in the signal and potential audible pops.

16.3. Pitch shifting through changes in the phase

Using the inverse DFT to reconstruct the signal is counterintuitive. The inverse DFT can produce only the frequencies of its components. If we keep track of the changes in phase, however, we will be able to construct any frequency.

In the example above, we calculated that the phase of the frequency 50 Hz changes by $\pi / 4$ for each 10 samples, whereas the phase of the frequency 75 Hz changes by $3 \pi / 4$ for each 10 samples. When calculating the inverse DFT, we use the DFT components, but also introduce a change in the phase of the DFT component frequency with each DFT segment.

The following figure shows the sum of overlapping segments of the frequency 50 Hz over the sampling frequency 2000 Hz. Each segment is of 40 samples and the segments overlap by 30 samples. That is, each next segment starts 10 samples later. The segments have carefully chosen phase, as explained below, and have been appropriately crossfaded, as discussed in the next section.

Figure 14. 75 Hz from overlapping 50 Hz segments

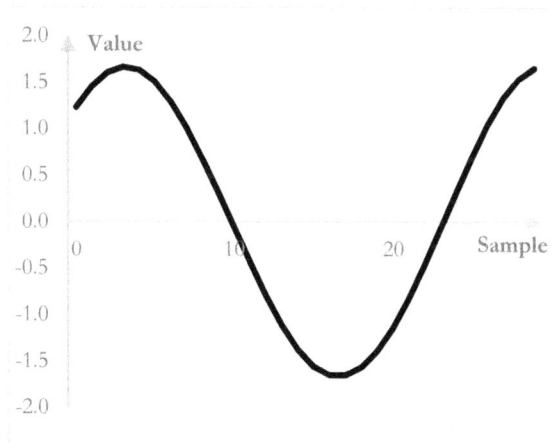

If we take several overlapping segments from a signal at 50 Hz and reconstruct this signal, but adjust the initial phase of each segment as if the signal is at 75 Hz, we get the signal on this figure. With the sampling rate 2000 Hz, the reconstructed signal completes a cycle in approximately 26 samples and so it is the 75 Hz signal.

Note that this function completes a cycle in approximately 26 samples. Since the sampling frequency is 2000 Hz, this is approximately the frequency 2000 / 26 = 75 Hz. In other words, we use overlapping segments of 50 Hz to create the frequency 75 Hz. We do so by choosing the appropriate initial phase for each segment.

The phase of the first segment is 0 (i.e., this is the function $\cos(2 \pi k 50 / 2000)$, where k is the sample). The phase of the second segment is $-3 \pi / 4$ (i.e., this is the function $\cos(3 \pi / 4 + 2 \pi$

k 50 / 2000)). The phase of the third segment is -6 π / 4, and so on. While we use segments of 50 Hz, with the offset of 10 samples per segment, the initial phases are those we expect of 75 Hz.

With the inverse DFT transform, pitch shifting is thus produced by the initial phases of the overlapping DFT segments at each DFT component.

16.4. Windowing

To crossfade, we use the *Hann window*. The Hann window is discussed in chapter 14 of volume 1. The amplitude flatness of the Hann window is 1 at various levels of overlap, including at 75% overlap. This means that crossfading with the Hann window produces no changes in the magnitude of the signal.

In addition to using the Hann window to crossfade the segments of the inverse DFT, we use the Hann window on the input to the forward DFT. This reduces spectral leakage and negates the change in amplitude in the graph above, where segments of 50 Hz with peak amplitude of 1 produce the frequency 75 Hz with peak amplitude of over 1.5.

16.5. Fast Fourier transform

A typical pitch shift implementation uses the *fast Fourier transform* (*FFT*). The FFT is discussed in chapter 13 of volume 1. It is not a different transform, but an algorithm to apply the DFT. The code snippets below do not use the FFT, as the goal is to present simpler code that is easy to understand.

16.6. Implementation of the pitch shift

The following are the member data of the pitch shift.

int FRAME – This is the size of the DFT. If the pitch shift effect uses the FFT, then **FRAME** must be a power of two. Appropriate DFT sizes are 1024, 2048, and so on. A DFT of 1024, for example, is sufficiently precise to pitch shift an acoustic guitar by four semitones up (an increase of the frequencies in the signal by a multiple of about 1.26). If the pitch shift uses the DFT, then the only requirement is that **FRAME** is divisible by **STEP** below.

int STEP – This is the step between two successive DFT segments in number of samples. To pitch shift the same acoustic guitar by a quarter of an octave, a step of 128 samples is sufficiently precise.

float m_pitch – This is the amount of pitch shift in the form of a multiple. Appropriate limits for the pitch are 0.5 and 2.0 (an octave down to an octave up). Since contemporary tuning splits an octave exponentially, a pitch shift of 4 semitones up, for example, is approximately $2^{4/12} \approx$ 1.26. Alternatively, a pitch shift of 1.26 is approximately $12 \log_2 1.26 \approx 4$.

double [] m_storeDry – This is the input to the DFT. The length of this array is the length of the DFT.

double [] m_storeWet0 – This is used to store the output of the DFT. The length of this array is equal to the number of samples in the buffer wet. Since each sample in wet is calculated from several overlapping inverse DFT computations, **m_storeWet0** accumulates the output of the DFT calculations until they are ready to be transferred to wet.

double [] m_storeWet1 – Since we need several overlapping DFT segments to calculate each output sample in **m_storeWet0**, some of the output of the inverse DFT calculations spills over **m_storeWet0**. It is stored in **m_storeWet1** to be accessed later.

int m_curByteDry – This counter keeps track of whether we have enough samples in **m_storeDry** to perform the DFT.

double [] m_magnitude – These are the magnitudes computed by the DFT for each of the frequencies of the DFT.

double [] m_newMagnitude – These are the magnitudes after the pitch shift that are used as input to the inverse DFT.

double [] m_phase – This array is used both to store the phase outputted by the DFT as well as the phase that is the input to the inverse DFT.

double [] m_previousPhase – This array stores the phase of the previous DFT segments. Comparing the phase produced by the previous DFT segment to the one produced by the current DFT segment allows us to calculate the true frequencies of the signal.

double [] m_accumulatePhase – This array accumulates the phase of the signal. This is necessary because the signal may change over time. If the signal did not change over time, we would be able to calculate the appropriate phase for the inverse DFT without accumulating the phase over the length of the entire signal.

double [] m_frequency – These are the "true" frequencies contained in the signal.

double [] m_newFrequency – These are the pitch shifted frequencies.

double [] m_hann – This is the Hann window used both in the forward DFT and the inverse DFT.

The constructor of the pitch shift is as follows.

Code 92. Pitch shift constructor

```
public Pitch()
{
    // Any value between, say, 0.5 and 2.0 (an octave down and an octave up) will
    // do
    m_pitch = 1.25F;

    // Most arrays can be of size FRAME / 2 rather than FRAME, since we are
    // working with a real valued signal (not a complex valued one) and the second
    // half of the DFT output is a duplicate of the first.  It is redundant
```

```
    m_storeDry = new double [FRAME];
    m_magnitude = new double [FRAME / 2];
    m_newMagnitude = new double [FRAME / 2];
    m_phase = new double [FRAME / 2];
    m_previousPhase = new double [FRAME / 2];
    m_accumulatePhase = new double [FRAME / 2];
    m_frequency = new double [FRAME / 2];
    m_newFrequency = new double [FRAME / 2];
    m_curByteDry = 0;

    // This is the Hann window
    m_hann = new double [FRAME];
    for(int i = 0; i < FRAME; i++)
        m_hann[i] = 0.5 * (1 - Math.cos(2D * Math.PI * i / (FRAME - 1)));

    // The size of these two arrays depends on the size of the incoming
    // and outgoing audio buffers.  These are set in the first call to
    // the apply function
    m_storeWet0 = null;
    m_storeWet1 = null;
}
```

The following code is executed at the beginning of playback.

Code 93. Pitch shift at the beginnning of playback

```
public void startPlay()
{
    // The first segment that is an input to the DFT will be mostly zero, except
    // for the last few samples (STEP number of samples).  As we progress through
    // overlapping DFT segments, the input to the DFT m_storeDry will be fully
    // populated with data
    m_curByteDry = FRAME - STEP;

    // Thus, we should zero m_storeDry out
    for(int i = 0; i < FRAME; i++)
        m_storeDry[i] = 0;

    // We can assume that the phase of all DFT frequencies in the first DFT
    // segment is zero (m_previousPhase)
    for(int i = 0; i < FRAME / 2; i++)
    {
        m_previousPhase[i] = 0;
        m_accumulatePhase[i] = 0;
    }
}
```

The implementation of the pitch shift is as follows.

Code 94. Pitch shift

```
public void apply(byte [] dry, byte [] wet, byte [] control, AudioFormat format,
    double time)
{
    // Initialize variables
    int blockAlign = (format.getChannels() * format.getSampleSizeInBits()) / 8;
    double sTo = 0F;

    // During the first call to this function, set up arrays.  We must do this
    // here rather than in startPlay, as we need blockAlign.  Technically, even
    // if the buffers are not null, we should also check their length, since the
    // user may have changed the size of the application audio buffers.  For
    // simplicity, we omit this here
    if (m_storeWet0 == null)
    {
        m_storeWet0 = new double [dry.length / blockAlign];
        m_storeWet1 = new double [dry.length / blockAlign];
        for(int i = 0; i < m_storeWet0.length; i++)
        {
            m_storeWet0[i] = 0;
            m_storeWet1[i] = 0;
        }
    }

    // This counter keeps track of the position of wet (pitch shifted)
    // audio in m_storeWet0
    int curWet = 0;

    // Accumulate samples from input (from dry), until we have enough samples
    // to perform a DFT, the pitch shift, and an inverse DFT.  Note the use "<="
    // in this loop, rather than "<".  This is to ensure that the DFT of the last
    // DFT segment is computed and used
    for(int i = 0; i <= dry.length; i += blockAlign, m_curByteDry++)
    {
        // If we have enough samples for a DFT
        if (m_curByteDry >= FRAME)
        {
            // Prepare this counter for the next DFT segment, by moving it back
            // by STEP.  This means that the DFT segments overlap by FRAME - STEP
            m_curByteDry = FRAME - STEP;

            // Calculate the forward DFT.  Since we are using the DFT on a real
            // signal, we only need half of the DFT output (hence, FRAME / 2).  The
            // remaining DFT output is a duplicate of this first half.  Note
            // the use of the Hann window on the DFT input
            for(int j = 0; j < FRAME / 2; j++)
            {
```

```
            double sines = 0;
            double cosines = 0;
            for(int k = 0; k < FRAME; k++)
            {
                sines += m_storeDry[k] * m_hann[k]
                    * Math.sin(2D * Math.PI * j * k / FRAME);
                cosines += m_storeDry[k] * m_hann[k]
                    * Math.cos(2D * Math.PI * j * k / FRAME);
            }

            // Calculate the magnitudes and phase that are the DFT output
            m_magnitude[j] = 2 * Math.sqrt(sines * sines + cosines * cosines)
                / FRAME;
            m_phase[j] = Math.atan2(sines, cosines);
        }

    // After the DFT, calculate the true frequencies present in the signal.
    // Since we know the number of samples between the start of this DFT
    // segment and the start of the previous DFT segment, we expect a
    // specific change in phase of the DFT frequency.  If the change is
    // different than the expected change, the true frequency contained by
    // the signal is not exactly the DFT frequency and we can compute what
    // it is.  We do so for each point of the DFT output
    double DFTphase = 0, phaseDifference = 0;
    for(int j = 0; j < FRAME / 2; j++)
    {
        // This is the expected change in the phase of the exact frequency of
        // the DFT component
        DFTphase = -2D * Math.PI * STEP * j / FRAME;

        // This is the change in the phase observed between two successive
        // DFT segments
        phaseDifference = m_phase[j] - m_previousPhase[j] - DFTphase;

        // Ensure that the change in phase is between -pi and pi by adding or
        // subtracting multiples of 2 pi as appropriate
        phaseDifference = phaseDifference - 2D * Math.PI
            * ((int) (phaseDifference / (2D * Math.PI)));
        if (phaseDifference > Math.PI)
            phaseDifference -= 2 * Math.PI;
        else if (phaseDifference < -Math.PI)
            phaseDifference += 2 * Math.PI;

        // Calculate the true frequency in the signal at each DFT component
        m_frequency[j] = ((float) j) * (format.getSampleRate() / FRAME)
            - phaseDifference * format.getSampleRate() / (2 * Math.PI * STEP);
    }
```

```
// Pitch shift by applying the pitch multiple to the frequencies and
// to the position of the magnitudes along the DFT components
for (int j = 0; j < FRAME / 2; j++)
{
   m_newMagnitude[j] = 0F;
   m_newFrequency[j] = 0F;
}

for (int j = 0; j < FRAME / 2; j++)
{
   int k = (int) ((float) j / m_pitch);
   if (k < FRAME / 2)
   {
      m_newMagnitude[j] += m_magnitude[k];
      m_newFrequency[j] = m_frequency[k] * m_pitch;
   }
}

// Compute the phase of the new frequency by doing the same thing as
// when computing the true frequency, but backwards
for(int j = 0; j < FRAME / 2; j++)
{
   phaseDifference = (((float) j) * (format.getSampleRate() / FRAME)
      - m_newFrequency[j]) * (2 * Math.PI * STEP) /
      format.getSampleRate();
   DFTphase = -2D * Math.PI * STEP * j / FRAME;
   phaseDifference += DFTphase;

   // Accumulate the newly computed phase for use in the next DFT
   // segment
   m_accumulatePhase[j] += phaseDifference;

   // Scale down the accumulated phase by multiples of 2 pi, since there
   // is no reason to use large numbers
   double sign = 1;
   if (m_accumulatePhase[j] < 0)
      sign = -1;
   m_accumulatePhase[j] = sign * (Math.abs(m_accumulatePhase[j])
      - 2D * Math.PI * ((int) (Math.abs(m_accumulatePhase[j])
      / (2 * Math.PI))));

   // Save the phase so that we can compute the true frequency in the
   // next DFT segment
   m_previousPhase[j] = m_phase[j];

   // The accumulated phase is the one that is used in the inverse DFT
   m_phase[j] = m_accumulatePhase[j];
}
```

```
// Compute the inverse DFT with the new magnitudes and phase.  If we use
// someone else's inverse DFT computation (e.g., by borrowing the code
// for the FFT), we may have to feed the inverse DFT with the real and
// and imaginary portions of the DFT components, which are respectively
// m_newMagnitude[j] * Math.cos(m_phase[j]) and m_newMagnitude[j] *
// Math.sin(m_phase[j]).  Here, we use the
// m_newMagnitude[j] * Math.cos(-m_phase[k] + …), which produces an
// equivalent result, as in chapter 5 of volume 1.  We may also have
// to use negative frequencies (set to zero).  Since we use our own
// inverse DFT implementation here, we do not have to do so
for(int j = 0; j < FRAME; j++, curWet++)
{
    sTo = 0;
    for(int k = 0; k < FRAME / 2; k++)
        sTo += m_newMagnitude[k] * Math.cos(-m_phase[k]
            + 2D * Math.PI * j * k / FRAME);

    // The inverse DFT is complete.  Adjust with the Hann window to
    // crossfade
    sTo *= m_hann[j];

    // Scale the output to the right sampling resolution
    sTo *= (double) Short.MAX_VALUE;
    sTo = Math.max(Math.min(sTo, Short.MAX_VALUE), Short.MIN_VALUE);

    // Store the result in the output buffer.  If the result falls out
    // of the output buffer, which will happen in the last
    // FRAME / STEP - 1 segments, store the result in the next output
    // buffer.  Note that we are adding sTo to what is already in the
    // output buffer.  This is the crossfading of overlapping DFT
    // segments
    if (curWet < m_storeWet0.length)
        m_storeWet0[curWet] += sTo;
    else
        m_storeWet1[curWet - m_storeWet0.length] += sTo;
}

// Adjust the start of where the result is stored in the output buffer
// for the next DFT segment
curWet -= (FRAME - STEP);

// Adjust the input to the forward DFT for the next DFT segment.  The
// first FRAME - STEP samples have already been prepared
System.arraycopy(m_storeDry, STEP, m_storeDry, 0, FRAME - STEP);
}
```

```
        // Prepare the input samples for the forward DFT.  The if statement is
        // needed, since the for loop above goes up to and including the length
        // the dry buffer
        if (i < dry.length)
        {
            m_storeDry[m_curByteDry] = (short)(((dry[i + 1] & 0xff) << 8)
                + (dry[i] & 0xff));
            m_storeDry[m_curByteDry] /= (double) Short.MAX_VALUE;
        }
    }

    // Shift the counter for input samples one down because of the "<="
    // condition in the for loop above
    m_curByteDry--;

    // Place the pitch shifted signal in the output
    for(int i = 0; i < m_storeWet0.length; i++)
    {
        wet[i * blockAlign] = (byte)((int) m_storeWet0[i] & 0xff);
        wet[i * blockAlign + 1] = (byte)((int) m_storeWet0[i] >>> 8 & 0xff);
    }

    // Copy the spillover of inverse DFT output into the output buffer
    // and zero out where that spillover is stored
    System.arraycopy(m_storeWet1, 0, m_storeWet0, 0, m_storeWet1.length);
    for(int i = 0; i < FRAME - STEP; i++)
        m_storeWet1[i] = 0;
}
```

16.7. Stretching or shrinking

The simplest way to implement stretching and shrinking of waves – without changing their pitch, of course – is to use the code for the pitch shifting with two minor changes. First, there should be no adjustments to magnitudes and phase after the DFT. **m_newMagnitude** and **m_newFrequency** will remain the same as **m_magnitude** and **m_frequency**. Second, the step **STEP** between overlapping inverse DFT segments should be different than the step between overlapping forward DFT segments. The wave is then stretched or shrunk when reconstructed with respectively a larger or a smaller step. Of course, the expected change in phase between overlapping segments **DFTphase** will be different in the forward and inverse DFT and hence the computed phase **m_phase** will change for the inverse DFT, when computed with the same code as in the pitch shift.

Chapter 17. Mixing and recording

A simple player is presented with the class **Mixer** in chapter 5. The mixer below is slightly more complex. It allows the mixing of information from several tracks, each with its own volume and pan.

17.1. Mixing several tracks

Below is a simplified version of the function **getData()** in the mixers of Orinj. This function gets audio data from all waves in a session and mixes these data into a single signal. In chapter 5, this function is implemented as a member function of the class **WaveFile**. Here, it does not belong to a single wave, but to the class responsible for playback and mixing.

The mixing below is not the most complex mixing possible. For example, it does not implement volume or pan envelopes – gradual changes in the volume and pan of tracks. It does not allow fast forwarding, rewinding, or looping. It does not handle side chaining and does not collect information for VU meters. It does, however, show two things, namely how the signals from several tracks should be mixed and how volume and pan values should be applied to tracks.

The following are selected member data of the encompassing class.

AudioBuffer m_bufferRead – This is the buffer used to get data from wave files on disk. It is of the class **AudioBuffer** described in chapter 5.

AudioBuffer m_buffersPlay – This is the mixed buffer sent to the output device for playback.

TrackMixer m_tracks – These are objects that represent individual tracks. Their implementation is not described here, but they will essentially be either wave files or mixers, if a track can have more than one wave file. Their job is solely to supply the audio data that should be mixed.

AudioFormat m_format – This is the format of audio data. This mixer assumes that all audio data have the same sampling rate and resolution.

Code 95. Mixing with getData

```
public boolean getData()
{
    // Here, depending on the implementation below, we may want to zero out
    // the playback buffer m_buffersPlay

    // Initialize some variables.  The first two are the input and output buffers.
    // In principle, we need only one buffer for reading and only one buffer to
    // be sent for playback.  The output device will copy the playback buffer into
    // its own buffer and release it for further use.  An exception maybe at the
    // beginning of playback, where we might want to pre-load the output device
    // with several playback buffers to make sure that, if there is much code
    // to be executed at the beginning of playback, playback starts smoothly
```

```
byte bufferFrom[] = m_bufferRead.getBuffer();
byte bufferTo[] = m_buffersPlay.getBuffer();
float volume = 0.0F;
float pan = 0.0F;
float sFrom = 0.0F;
float sTo = 0.0F;
int blockAlign = (m_format.getChannels() * m_format.getSampleSizeInBits())
    / 8;


// This class is not described here, but it is the class that obtains audio
// data from a single track and mixes it into one signal for that track.
// If a track can contain only one wave file, then this class can be
// implemented as the class WaveFile of chapter 5.  If a track can contain
// more than one wave, then this class can be implemented similarly to the
// mixer described here
TrackMixer trackMixer = null;

// Go through each track
for(int i = 0; i < m_tracks.length; i++)
{
   // Assume that the buffer has no data.  The class AudioBuffer of chapter 5
   // can be expanded to include a boolean flag for whether the buffer has
   // data.  A buffer can be empty for several reasons, including if the track
   // is empty, if the track is muted, or if the waves in the track are not at
   // the position that the buffer is trying to read.  In principle, this flag
   // is not absolutely necessary, as the buffer can simply be filled with
   // zeroes if there is no audio data.  However, the flag is a good idea, as
   // it allows us to skip computations with zero audio data, which saves
   // computational time
   m_bufferRead.setHasData(false);

   // This is the mixer for the current track
   trackMixer = m_tracks[i];

   // Obtain audio data from the track.  Again, if a track can have only
   // one wave, this getData function can be implemented similarly to the
   // getData function of the class WaveFile in chapter 5.  If a track can
   // have more than one wave, then this getData function can be implemented
   // similarly to this function here.  Here, we expect that the function will
   // populate m_bufferRead with data and will need to know the audio format.
   // getData will return false, if there was an error in obtaining the audio
   // data and true otherwise.  This function could have other arguments, such
   // as the beginning of playback if the user may start playback not at time
   // zero, but at some other time in the session
   if(!trackMixer.getData(m_bufferRead, m_format))
      return false;

   // If the buffer has some audio data
```

```
if(m_bufferRead.getHasData())
{
    // Get the track volume.  For the purposes of this code snippet, assume
    // that each track is equipped with a volume value in decibels (say,
    // between -40 dB and 40 dB)
    volume = (float)Math.pow(10D, trackMixer.getVolume() / 20F);

    // Get the track pan.  For the purposes of this code snipped, assume
    // that each track is equipped with a pan value between -1 (far left)
    // and 1 (far right)
    pan = trackMixer.getPan();

    // Go through the audio data in the buffer
    for(int j = 0; j < m_bufferRead.getSize(); j +=
        m_format.getSampleSizeInBits() / 8)
    {
        // Get the value of the sample that mixes all previous tracks
        sTo = (short)(((bufferTo[j + 1] & 0xff) << 8)
            + (bufferTo[j] & 0xff));

        // Get the value of the sample from the current track
        sFrom = (short)(((bufferFrom[j + 1] & 0xff) << 8)
            + (bufferFrom[j] & 0xff));

        // Apply the track volume
        sFrom = volume * sFrom;

        // Apply the track pan.  Assume that each tack produces stereo
        // two channel data.  The first two bytes (j % blockAlign == 0) are
        // for the left channel.  The last two bytes are for the right
        // channel.  This only works for two channels and not more
        if(j % blockAlign == 0)
            sFrom = (1.0F - pan) * sFrom;
        else
            sFrom = (1.0F + pan) * sFrom;

        // Add the current track to the mix of all previous tracks
        sTo = Math.min(Math.max(sTo + sFrom, Short.MIN_VALUE),
            Short.MAX_VALUE);

        // Store the output
        bufferTo[j] = (byte)((int)sTo & 0xff);
        bufferTo[j + 1] = (byte)((int)sTo >>> 8 & 0xff);
    }
}
}
```

```
    // We adjust the buffer start time.  The buffer start time can be used to tell
    // tracks what part of the audio data to get.  The line of code below simply
    // moves the buffer to the next audio segment.  We can move the buffer
    // elsewhere if we want to implement fast forwarding or rewinding.  We can
    // also check whether the buffer exceeds the end of the session, in which
    // case playback should stop
    m_bufferRead.setTimeStart(m_bufferRead.getTimeStart() +
        m_bufferRead.getTimeLength());

    return true;
}
```

17.2. Recording

The important thing to remember about recording is that it should allow for simultaneous playback. The user may need to listen to other tracks when recording a new one. Recording is therefore typically implemented as part of the playback thread. If recording is a separate thread, then care must be taken so that the two threads start at (approximately) the same time, although they do not necessarily need to be synchronized.

The following code is executed at the beginning of recording and simultaneous playback.

Code 96. Start recording

```
public boolean startPlay(double timeEnd)
{
    // Here, we might want to prepare tracks for playback, check that the buffers
    // have been set up with the correct sizes and/or start times, and obtain some
    // audio data that we can feed into the output sound device

    ...

    // Check if we are recording.  Assume that m_recording is a flag set by the
    // user (e.g., when pressing the record button of the application)
    if(m_recording)
    {
        // Go through each track
        for(int j = 0; j < m_tracks.length; j++)
        {
            // Prepare the track for recording, if the user chose to record to
            // this track (e.g., make sure that the track has a wave file to write
            // to)

            ...

            // Try to obtain an input device
            Mixer mixer = null;
            try
```

```
   {
      mixer = AudioSystem.getMixer(m_tracks[j].getInputDevice().getInfo());
   }
   catch (IllegalArgumentException e)
   {
      // Display an error, such as
      // JOptionPane.showMessageDialog(MyMainFrm.m_frame,
      //    "Detected errors when attempting to record", "Error",
      //    JOptionPane.ERROR_MESSAGE);
      return false;
   }
   catch (SecurityException e)
   {
      // Display an error
      return false;
   }

   if (mixer == null)
   {
      // Display an error
      return false;
   }

   // Try to make the input device get the input line with the right
   // audio format
   DataLine.Info info = new DataLine.Info(TargetDataLine.class, m_format);
   if (!mixer.isLineSupported(info))
   {
      // Display an error
      return false;
   }

   try
   {
      m_mike = (TargetDataLine) mixer.getLine(info);
      if (m_mike == null)
      {
         // Display an error
         return false;
      }
      if (m_mike.isActive())
         m_mike.stop();
      m_mike.open(m_format, m_mike.getBufferSize());
   }
   catch (LineUnavailableException e)
   {
      // Display an error
```

```
                return false;
        }
        catch (IllegalArgumentException e)
        {
            // Display an error
            return false;
        }
    }
}

// We are now ready to play and record.  Get an output device and the output
// line with the right format.  Start playback in a separate thread here or
// elsewhere

...

return true;
}
```

The following is the recording and simultaneous playback thread.

Code 97. Record

```
public void run()
{
    // Start the input device
    if(m_recording && m_mike != null)
        m_mike.start();

    // Start the output device.  We start the input and output device together and
    // at the beginning of the playback thread so that there is both little
    // difference in latencies and little latency
    m_sourceLine.start();

    // Here, we could start recording in a new thread.  A new thread is not
    // needed if the playback audio buffers and the recording audio buffers are of
    // the same size.  We assume that this is the case here and process the
    // recorded data below, rather than in a new thread.  If this is not the case
    // and the buffers are of different sizes, it is best that the recording is
    // independent of the playback and is placed in a separate thread

    // Write whatever buffers are already ready to output.  This fills the
    // output device buffers with enough data, so that it can play while we
    // perform other operations
    for(int j = 0; j < m_buffersPlay.length - 1; j++)
        m_sourceLine.write(m_buffersPlay[j].getBuffer(), 0,
            m_buffersPlay[j].getActualSize());

    // Continue playback and recording until the user stops it.  Here, assume
```

```
// that if the user stops playback and recording, m_playing will be set to
// false somewhere in the application, outside of this thread
while (m_playing)
{
    // Use Thread.sleep(long) here if the user paused playback.  Otherwise,
    // continue with the operations below

    // Get data for playback from the hard disk.  An example of the function
    // getData() is above
    if(!getData() && (!m_recording))
    {
        // Display an error
        return;
    }

    // Send playback data to the output device.
    m_sourceLine.write(m_buffersPlay.getBuffer(), 0,
        m_buffersPlay.getActualSize());

    // If we are recording
    if(m_recording && m_mike != null)
    {
        // If enough time has passed and the available recorded data are larger
        // than the size of the recording buffer
        while (m_mike.available() >= m_buffersRecord.getSize())
        {
            // Get the recorded data
            m_bytesRecorded = m_mike.read(m_buffersRecord.getBuffer(), 0,
                m_buffersRecord.getSize());

            // Write the recorded data to the hard disk.  The implementation of
            // writeData() should be simple and should simply write the buffer
            // to the wave file
            if (!writeData())
            {
                // Display an error
                return;
            }
        }
    }

    // Here, we may want to stop playback if we are not recording and
    // the playback time exceeds the end of the wave or session
    ...
}

// Stop playback and close the source line
```

...

```
// Stop recording
if(m_mike != null)
{
    // Stop the mike
    m_mike.stop();

    // Check if there is any remaining available data in the mike, get these
    // data, and write them to the hard disk
    int available = m_mike.available();
    while (available > 0)
    {
        m_bytesRecorded = m_mike.read(m_buffersRecord.getBuffer(), 0,
            Math.min(available, m_buffersRecord.getSize()));
        available -= m_bytesRecorded;
        if(m_bytesRecorded == 0)
            break;
        if(!writeData())
        {
            // Display an error
            return;
        }
    }

    // Close the mike
    m_mike.drain();
    m_mike.close();

    m_recording = false;
}

// Here, perform any actions that should be performed for the recorded tracks
// (e.g., close recorded wave files and perhaps open them for reading only for
// future playback)
...
}
```

Appendix A. Other wave chunks

A.1. Cue chunk

The cue chunk specifies cues or markers in the wave file, such as the beginning of a verse or the end of a solo.

Figure 15. Structure of the cue chunk in a wave file

Byte sequence description	Length in bytes	Starts at byte in the chunk	Value
chunk ID	4	0x00	The ASCII character string "cue " (note the space)
size	4	0x04	The size of the cue chunk (number of bytes) less 8 (less the "chunk ID" and the "size")
number of cue points	4	0x08	The number of cue points in the list of cue points that follows
data	various	0x0C	A list of cue points. Each cue point uses 24 bytes of data

The cue chunk is optional, but, if it exists, at most one cue chunk per file is allowed. Thus, the single cue chunk should contain all cue points in the file.

A cue point uses 24 bytes of data as follows.

Figure 16. Structure of a cue point in a cue chunk

Byte sequence description	Length in bytes	Starts at byte in the cue point	Value
ID	4	0x00	A unique number for the point used by other chunks to identify the cue point. For example, a playlist chunk creates a playlist by referring to cue points, which themselves define points somewhere in the file
position	4	0x04	If there is no playlist chunk, this value is zero. If there is a playlist chunk, this value is the sample at which the cue point should occur

data chunk ID	4	0x08	Either "data" or "slnt" depending on whether the cue occurs in a data chunk or in a silent chunk
chunk start	4	0x0C	The position of the start of the data chunk that contains the cue point. If there is a wave list chunk, this value is the byte position of the chunk that contains the cue. If there is no wave list chunk, there is only one data chunk in the file and this value is zero
block start	4	0x10	The byte position of the cue in the "data" or "slnt" chunk. If this is an uncompressed PCM file, this is counted from the beginning of the chunk's data. If this is a compressed file, the byte position can be counted from the last byte from which one can start decompressing to find the cue
sample start	4	0x14	The position of the cue in number of bytes from the start of the block

A.2. Fact chunk

A fact chunk contains the following information.

Figure 17. Structure of a fact chunk in a wave file

Byte sequence description	Length in bytes	Starts at byte in the chunk	Value
chunk ID	4	0x00	The ASCII character string "fact"
size	4	0x04	The size of the fact chunk (number of bytes) less 8 (less the "chunk ID" and the "size")
data	4	0x08	Various information about the contents of the file, depending on the compression code

Fact chunks exist in all wave files that are compressed or that have a wave list chunk. A fact chunk is not required in an uncompressed PCM file that does not have a wave list chunk.

According to the fact chunk's initial specification, the data portion of the fact chunk will contain only one 4-byte number that specifies the number of samples in the data chunk of the Wave file. This number, when combined with the samples per second value in the format chunk of the wave file, can be used to compute the length of the audio data in seconds.

A.3. Instrument chunk

When a wave file is used as wave samples in a Musical Instrument Digital Interface (MIDI) synthesizer, the instrument chunk helps the MIDI synthesizer define the sample pitch and relative volume of the samples. The instrument chunk has the following structure.

Figure 18. Structure of an instrument chunk in a wave file

Byte sequence description	Length in bytes	Starts at byte in the chunk	Value
chunk ID	4	0x00	The ASCII character string "inst"
size	4	0x04	The size of the chunk less 8 (less the "chunk ID" and the "size")
unshifted note	1	0x08	The MIDI note that corresponds to the original (unshifted) pitch of the sample. This value is between 0 to 127, as in a standard MIDI Note On message
fine tuning	1	0x09	Fine tuning of the pitch in cents. Values are between -50 to 50
gain	1	0x0A	The volume setting (suggested) for the sample in decibels
low note	1	0x0B	The lowest usable MIDI note for the sample (suggested). This value is between 0 and 127
high note	1	0x0C	The highest usable MIDI note for the sample (suggested). This value is between 0 and 127
low velocity	1	0x0D	The lowest usable MIDI velocity for the sample (suggested). This value is between 0 and 127
high velocity	1	0x0E	The highest usable MIDI velocity for the sample (suggested). This value is between 0 and 127

A.4. List chunk

A list chunk defines a list of sub-chunks and has the following format.

Figure 19. Structure of a list chunk in a wave file

Byte sequence description	Length in bytes	Starts with byte in the chunk	Value
chunk ID	4	0x00	The ASCII character string "list"
size	4	0x04	The size of the sub-chunk less 8 (less the "chunk ID" and the "size")
list type ID	4	0x08	Various ASCII character strings. Two common types are "adtl" (associated data list) and "info" (text information about copyright, authorship, etc.)
data	various	0x0C	Depends on the list type ID

An associated data list, for example, would define text labels and names for cue points. An associated data list typically uses label sub-chunks, note sub-chunks, and labeled text sub-chunks.

A label sub-chunk and a note sub-chunk have the same structure.

Figure 20. Structure of a label or a note sub-chunk

Byte sequence description	Length in bytes	Starts with byte in the sub-chunk	Value
sub-chunk ID	4	0x00	The ASCII character string "labl" or "note"
size	4	0x04	The size of the chunk less 8 (less the "sub-chunk ID" and the "size")
cue point ID	4	0x08	The ID of the relevant cue point (see cue chunk above)
data	various	0x0C	Some ASCII text

The ASCII text is null terminated and must be padded, if not word aligned.

The labeled text sub-chunk associates some portion of the audio data with text and serves as a marker. This sub-chunk has the following format.

Figure 21. Structure of a labeled text sub-chunk

Byte sequence description	Length in bytes	Starts at byte in the sub-chunk	Value

sub-chunk ID	4	0x00	The ASCII character string "ltxt"
size	4	0x04	The size of the chunk less 8 (less the "sub-chunk ID" and the "size")
cue point ID	4	0x08	The ID of the relevant cue point
sample length	4	0x0C	The number of samples in the segment of audio data described by this sub-chunk
purpose ID	4	0x10	The purpose of the text. Common IDs are "scrp" for script and "capt" for closed captioning
country	2	0x14	The country of the text
language	2	0x16	The language of the text
dialect	2	0x18	The dialect of the text
code page	2	0x1A	The code page for the text
data	various	0x1C	Some ASCII text

The labeled text sub-chunk is always a part of an associated data list chunk.

When a list chunk carries the list type ID "info", the list contains information about the copyright, author, engineer of the file, and other similar text. The data of the list chunk are organized as follows.

> Info ID (4 byte ASCII text) for information 1
> Size of text 1
> Text 1
> Info ID (4 byte ASCII text) for information 2
> Size of text 2
> Text 2
> ...

Common info IDs in Wave files are as follows.

Figure 22. Common info IDs

Info ID	The corresponding text describes
"iart"	The artist of the original subject of the file
"icms"	The name of the person or organization that commissioned the original subject of the file
"icmt"	General comments about the file or its subject
"icop"	Copyright information about the file (e.g., "Copyright Some Company 2011")
"icrd"	The date the subject of the file was created (creation date)

"ieng"	The name of the engineer who worked on the file
"ignr"	The genre of the subject
"ikey"	A list of keywords for the file or its subject
"imed"	Medium for the original subject of the file
"inam"	Title of the subject of the file (name)
"iprd"	Name of the title the subject was originally intended for
"isbj"	Description of the contents of the file (subject)
"isft"	Name of the software package used to create the file
"isrc"	The name of the person or organization that supplied the original subject of the file
"isrf"	The original form of the material that was digitized (source form)
"itch"	The name of the technician who digitized the subject file

All text must be word aligned.

A.5. Playlist chunk

A playlist chunk contains the following information.

Figure 23. Structure of a playlist chunk in a wave file

Byte sequence description	Length in bytes	Starts at byte in the chunk	Value
chunk ID	4	0x00	The ASCII character string "plst"
size	4	0x04	The size of the playlist chunk (number of bytes) less 8 (less the "chunk ID" and the "size"). Since each segment has 12 bytes, this is also equal to the number of segments multiplied by 12
number of segments	4	0x08	The number of segments that follow
data	various	0x0C	A list of segments

The playlist defines a series of cue points (see cue chunk above), which refer to some segments of the file, defines in what order the segments should be played, how long each segment is and how many times it should be repeated during playback.

Each segment in the playlist has the following structure.

Figure 24. Structure of a segment of a playlist chunk

Byte sequence description	Length in bytes	Starts at byte in the segment	Value
cue point ID	4	0x00	The ID of the cue point in the cue chunk
segment length	4	0x04	The length of the segment in number of samples to play from the sample defined by the cue point
repeats	4	0x08	The number of times the segment should be repeated

A.6. Sample chunk

The sample chunk allows a Musical Instrument Digital Interface (MIDI) sampler to use the Wave file as a collection of samples.

Figure 25. Structure of a sample chunk in a wave file

Byte sequence description	Length in bytes	Starts at byte in the chunk	Value
chunk ID	4	0x00	The ASCII character string "smpl"
size	4	0x04	The size of the chunk less 8 (less the "chunk ID" and the "size")
manufacturer	4	0x08	The MIDI Manufacturers Association manufacturer code (typically used in the MIDI System Exclusive message). A value of zero implies that there is no specific manufacturer. The first byte of the four bytes specifies the number of bytes in the manufacturer code that are relevant (1 or 3). For example, Roland would be specified as 0x01000041 (0x41), whereas Microsoft would be 0x03000041 (0x00 0x00 0x41)
product	4	0x0C	The product / model ID of the target device, specific to the manufacturer. A value of zero means no specific product

sample period	4	0x10	The period of one sample in nanoseconds. For example, at the sampling rate 44.1 kHz the size of one sample is (1 / 44100) * 1,000,000,000 = 22675 nanoseconds = 0x00005893
MIDI unity note	4	0x14	The MIDI note that will play when this sample is played at its current pitch. The values are between 0 and 127, as in the MIDI Note On message)
MIDI pitch fraction	4	0x18	The fraction of a semitone up from the specified note. For example, one-half semitone is 50 cents and will be specified as 0x80
SMPTE format	4	0x1C	The SMPTE format. Possible values are 0, 24, 25, 29, and 30, which are the frames per second in the MIDI time division
SMPTE offset	4	0x20	Specifies a time offset for the sample, if the sample should start at a later time and not immediately. The first byte of this value specifies the number of hours and is in between -23 and 23. The second byte is the number of minutes and is between 0 and 59. The third byte is the number of seconds (0 to 59). The last byte is the number of frames and is between 0 and the frames specified by the SMPTE format. For example, if the SMPTE format is 24, then the number of frames is between 0 and 23
number of sample loops	4	0x24	Specifies the number of sample loops that are contained in this chunk's data
sample data	4	0x28	The number of bytes of optional sampler specific data that follow the sample loops. If there are no such data, the number of bytes is zero
sample loops	various	0x2C	The sample loops
sampler specific data	various	various	The sampler specific data are optional

A sample loop has the following structure.

Figure 26. Structure of a sample loop

Byte sequence description	Length in bytes	Starts at byte in the sub-chunk	Value
ID	4	0x00	A unique ID of the loop, which could be a cue point (see cue chunk above)
type	4	0x04	The loop type. A type of 0 means normal forward looping type. A value of 1 means alternating (forward and backward) looping type. A value of 2 means backward looping type. The values 3-31 are reserved for future standard types. The values 32 and above are sampler / manufacturer specific types
start	4	0x08	The start point of the loop in samples
end	4	0x0C	The end point of the loop in samples. The end sample is also played
fraction	4	0x10	The resolution at which this loop should be finetuned. A value of zero means current resolution. A value of 50 cents (0x80) means ½ sample
number of times to play the loop	4	0x14	The number of times to play the loop. A value of zero means infinitely. In a MIDI sampler that may mean infinite sustain

A.7. Silent chunk

A silent chunk has the following structure.

Figure 27. Structure of a silent chunk in a wave file

Byte sequence description	Length in bytes	Starts at byte in the chunk	Value

chunk ID	4	0x00	The ASCII character string "slnt"
size	4	0x04	The value 4. This value is the size of the silent chunk (number of bytes) less 8 (less the "chunk ID" and the "size")
data	4	0x08	The number of samples through which playback should be silent

Playback will halt at the last sample value, which could be silence, or could be anything else. To avoid pops and clicks, the preceding data chunk should end in silence.

A.8. Wave list chunk

A wave list chunk contains a sequence of alternating silent chunks and data chunks.

Figure 28. Structure of a wave list chunk in a wave file

Byte sequence description	Length in bytes	Starts at byte in the chunk	Value
chunk ID	4	0x00	The ASCII character string "wavl"
size	4	0x04	The size of the chunk less 8 (less the "chunk ID" and the "size")
data	various	0x08	A sequence of alternating silent and data chunks

Index

www.ingramcontent.com/pod-product-compliance
Lightning Source LLC
Chambersburg PA
CBHW061417210326
41598CB00035B/6250